从LDP到SR：
MPLS网络技术实战

马海龙 伊 鹏 江逸茗
张 鹏 王 亮 谢记超 ◎ 编著

人民邮电出版社
北京

图书在版编目（CIP）数据

从LDP到SR：MPLS网络技术实战 / 马海龙等编著
. -- 北京：人民邮电出版社，2024.3
ISBN 978-7-115-62717-9

Ⅰ．①从… Ⅱ．①马… Ⅲ．①宽带通信系统—综合业务通信网 Ⅳ．①TN915.142

中国国家版本馆CIP数据核字(2023)第181314号

内 容 提 要

本书从技术实践入手，以 MPLS 技术为主线，讲解了 LDP 和基于 MPLS 的 Segment Routing 机制。首先，阐述了基于 EVE-NG 的实验环境构建、MPLS 的技术原理，并结合静态标签绑定方法讲述了 MPLS 数据转发路径的验证原理和方法；其次，介绍了 LDP 原理，借助实例详细阐述了 LDP 会话机制、标签分发、表项生成等过程，以及 LDP 配置和 MPLS 路径验证方法；最后，分别阐述了通过拓展 OSPF、IS-IS 协议及 BGP 功能以支持 Segment Routing 标签分发的机制及配置实践，并进一步介绍了基于 Segment Routing 的快速重路由技术和防微环路技术。

本书适用于计算机科学与技术、信息与通信工程、网络空间安全等相关学科专业的研究生，也可供网络技术人员、网络工程人员参考。

◆ 编　著　马海龙　伊　鹏　江逸茗
　　　　　　张　鹏　王　亮　谢记超
　责任编辑　李彩珊
　责任印制　马振武

◆ 人民邮电出版社出版发行　北京市丰台区成寿寺路 11 号
　邮编　100164　电子邮件　315@ptpress.com.cn
　网址　https://www.ptpress.com.cn
　固安县铭成印刷有限公司印刷

◆ 开本：720×960　1/16
　印张：19　　　　　　　　　　2024 年 3 月第 1 版
　字数：341 千字　　　　　　　2024 年 3 月河北第 1 次印刷

定价：189.80 元

读者服务热线：(010)81055493　印装质量热线：(010)81055316
反盗版热线：(010)81055315
广告经营许可证：京东市监广登字 20170147 号

前言

随着云计算、大数据、人工智能、物联网等技术的快速成熟，各类业务对网络带宽、性能、可靠性等指标的要求不断攀升。Segment Routing 技术以其灵活的可编程能力和良好的拓展能力，逐渐成为各类新兴网络应用的有力支撑技术，得到了网络运营者的青睐，并在各种网络场景中开展部署应用。MPLS 网络技术虽很早被提出并应用于服务质量保证，但控制协议的复杂度，影响了其大规模应用。Segment Routing 技术重用了 MPLS 转发机制，赋予标签新的内涵，使得 MPLS 技术得到了更大范围的部署。Segment Routing 技术兼容现有 MPLS 机制，且具有很好的互操作性，使得现有网络设备仅仅通过控制软件的升级就可以实现对 Segment Routing 的支持。因此，MPLS 网络技术又迎来了新的发展机遇。

为帮助渴望快速了解 MPLS 网络技术和掌握 LDP、Segment Routing 路由协议的网络工程师、网络管理员和相关专业学生，作者着眼"技术原理+实战赋能"的思路撰写了本书。全书内容可分为 3 个部分，分别是实验环境构建及 MPLS 技术原理、LDP 机制及实践、Segment Routing 核心控制协议机制及快速重路由技术，主要内容概要如下。

第一部分介绍了 MPLS 网络基础知识，包括 MPLS 的原理、架构、组成部分、标签交换等基础概念，并结合静态标签绑定方法讲述了 MPLS 数据转发路径的验证原理和方法。MPLS 技术的核心是标签交换，它能为数据包分配一个标签，然后在 MPLS 网络中通过标签来快速转发，从而提供了高效的数据传输能力。这种基于静态标签绑定的方式，可使我们快速了解 MPLS 数据包的转发原理，从而掌握 MPLS 数据转发机制。

第二部分围绕LDP机制及实践展开，我们深入剖析了LDP的原理、实现和配置方法，并结合实战场景介绍了LDP的路径建立过程。LDP通过自动化的方法，为MPLS数据转发提供了高效的标签生成机制。在书中，我们分别介绍了LDP会话机制、标签分发、表项生成等过程，并实例化地阐述了LDP网络的配置和MPLS路径验证方法。

第三部分则以Segment Routing为主线，详细讲解了如何通过扩展OSPF、IS-IS及BGP机制的相应功能来支持Segment Routing所需的标签分配和分发，帮助读者更好地掌握Segment Routing控制面的实现原理。进一步介绍了基本的LFA路径、R-LFA路径等重路由技术以及基于Segment Routing的TI-LFA路径重路由和防微环路技术。结合实验实例，帮助读者在理解Segment Routing控制协议的基础上，了解和掌握Segment Routing技术的典型应用。

本书的特点在于融合了理论与实践，通过"技术原理+实战赋能"的方式，帮助读者快速理解MPLS网络的实际应用和技术细节，掌握LDP和Segment Routing在实际环境中的配置、调试等技术。本书核心内容涵盖了以MPLS技术为核心的控制协议机制及与协议配置相关的实际操作过程，对于从事网络规划、设计、建设、运维等方面的网络从业者都有参考价值。

作者贡献方面，马海龙教授负责本书总体行文思路组织、章节安排，独立编写了第5章和第8章；伊鹏研究员独立编写了第3章、第4章；江逸茗副研究员完成了第7章编写，参与第6章编写；张鹏副研究员独立完成了第2章编写；王亮助理研究员完成了第1章编写，并参与全书统编和文字整理；谢记超助理研究员完成全书编排、图表制作等；罗伟工程师完成了第6章编写。

最后，我们衷心感谢出版社的鼎力支持和责任编辑的悉心帮助，让本书能够如期出版。如果您在阅读过程中有任何问题或建议，欢迎您随时与我们交流。我们期待着与您分享更多有关MPLS网络技术的实用经验和技巧。

作　者
2023年5月于郑州

目 录

第 1 章 绪论 ·· 1

1.1 引言 ··· 1
1.2 MPLS 与 Segment 技术起源与发展 ··· 2
1.2.1 MPLS 技术起源、发展与不足 ·· 2
1.2.2 Segment Routing 技术起源与发展 ·· 3
1.3 Segment Routing 技术基础原理 ·· 4
1.3.1 基本概念 ·· 4
1.3.2 Segment Routing 技术原理 ·· 6
1.3.3 Segment Routing 架构 ·· 6
1.4 Segment Routing 应用前景 ·· 9
1.4.1 Segment Routing 用于现有 MPLS 网络升级 ···························· 9
1.4.2 Segment Routing 用于数据中心 ··· 9
1.4.3 Segment Routing 用于大规模互联 ······································· 10
1.4.4 承载网络 ··· 10
1.4.5 SD-WAN 构建 ·· 10
1.5 本书内容组织 ·· 10

第 2 章 基于 EVE-NG 的实验环境构建 ·· 12

2.1 EVE-NG 概述 ··· 12

2.1.1 EVE-NG 的用途 ··· 13
2.1.2 EVE-NG 安装 ··· 14
2.2 实验仿真环境构建 ··· 18
2.2.1 客户端集成软件包的安装 ························· 18
2.2.2 虚拟节点镜像管理 ····························· 20
2.2.3 Lab 管理 ······································ 22
2.3 路由器配置管理 ·· 34
2.3.1 命令行工作模式 ································ 34
2.3.2 配置与保存 ···································· 35
2.4 CPSPT 五步实验法 ·· 38
2.5 实验约定 ·· 39

第 3 章 MPLS 技术原理与实战 ·· 41

3.1 MPLS 原理 ·· 41
3.2 MPLS 静态配置 ·· 43
3.2.1 实验环境 ······································ 44
3.2.2 协议配置 ······································ 44
3.2.3 使能 MPLS 转发功能 ·························· 47
3.3 MPLS 数据转发 ·· 48
3.3.1 MPLS Echo 模式 ································ 48
3.3.2 MPLS Echo 报文封装机制 ························ 50
3.3.3 MPLS ping ····································· 53
3.3.4 MPLS traceroute ································ 58
3.4 MPLS TTL 传播机制 ······································ 61

第 4 章 LDP 技术原理与实战 ·· 66

4.1 LDP 简介 ·· 66
4.2 LDP 机制 ·· 66
4.2.1 实验环境 ······································ 66

    4.2.2　协议配置 ·· 67
    4.2.3　LDP 交互 ··· 72
  4.3　转发路径验证 ·· 83
    4.3.1　连通性验证 ·· 83
    4.3.2　标签交换路径验证 ·· 87
    4.3.3　MPLS 分组的 TTL 超时响应 ·· 88

第 5 章　OSPF-SR 原理与实战 ·· 91
  5.1　OSPF-SR 原理 ··· 91
    5.1.1　基于 IGP 分发 SR 标签的优势 ·· 91
    5.1.2　MPLS 标签与 Segment ID ·· 92
    5.1.3　OSPF 链路状态信息与 SID 的关联 ·· 93
    5.1.4　SID 信息全域洪泛 ··· 94
    5.1.5　MPLS 转发表项生成原理 ·· 95
  5.2　域内 SR 实验 ··· 100
    5.2.1　实验环境 ·· 100
    5.2.2　协议配置 ·· 101
    5.2.3　协议交互过程 ·· 101
    5.2.4　数据转发验证 ·· 111
  5.3　域间 SR 实验 ··· 123
    5.3.1　实验环境 ·· 123
    5.3.2　协议交互 ·· 124
    5.3.3　数据转发验证 ·· 129
  5.4　外部 SR 实验 ··· 130
    5.4.1　实验环境 ·· 130
    5.4.2　协议配置 ·· 131
    5.4.3　协议交互 ·· 133
    5.4.4　数据转发验证 ·· 135

第 6 章 IS-IS SR 原理与实战 ·········· 136

6.1 IS-IS SR 原理 ·········· 136
6.1.1 IS-IS 链路状态信息与 SID 的关联 ·········· 136
6.1.2 SID 信息的洪泛方式 ·········· 137

6.2 域内 SR 实验 ·········· 137
6.2.1 实验环境 ·········· 137
6.2.2 协议配置 ·········· 138
6.2.3 协议交互过程 ·········· 139
6.2.4 数据转发验证 ·········· 148

6.3 域间 SR 实验 ·········· 159
6.3.1 实验环境 ·········· 159
6.3.2 协议配置 ·········· 160
6.3.3 协议交互 ·········· 163
6.3.4 数据转发验证 ·········· 170

6.4 外部 SR 实验 ·········· 171
6.4.1 实验环境 ·········· 171
6.4.2 协议配置 ·········· 172
6.4.3 协议交互 ·········· 173
6.4.4 数据转发验证 ·········· 175

第 7 章 SR BGP 原理与实战 ·········· 177

7.1 SR BGP 基本原理 ·········· 177
7.1.1 基本概念 ·········· 177
7.1.2 基于 BGP-LU 的 PrefixSID 通告机制 ·········· 177
7.1.3 基于 BGP-LS 的 PeeringSID 通告机制 ·········· 179

7.2 基于 BGP-LU 的标签分发 ·········· 180
7.2.1 实验环境 ·········· 180
7.2.2 基本配置 ·········· 181

7.2.3 BGP-LU 机制 ·········· 184
7.2.4 数据面验证 ·········· 198
7.3 基于 BGP-LU 的 PrefixSID 分发 ·········· 202
7.3.1 实验环境 ·········· 202
7.3.2 BGP-LU 标签与 IGP-SR 标签的不一致性 ·········· 203
7.3.3 BGP-SR 的标签分配 ·········· 207
7.3.4 BGP-SR 标签与 BGP-LU 标签的一致性 ·········· 209
7.3.5 数据转发面验证 ·········· 211
7.4 BGP EPE ·········· 215
7.4.1 实验环境 ·········· 215
7.4.2 EPE 配置与通告 ·········· 217
7.5 混合多自治域环境下的混合 SID 传递实验 ·········· 225
7.5.1 实验环境 ·········· 225
7.5.2 缺省策略条件下的路由与数据转发 ·········· 226
7.5.3 基于 BGP PrefixSID 实现跨 AS 引流 ·········· 226
7.5.4 基于 BGP PeeringSID 实现跨域间链路引流 ·········· 228
7.5.5 基于 IGP SID 实现跨域内引流 ·········· 228

第 8 章 基于 SR 的 FRR 技术原理与实战 ·········· 238

8.1 FRR 技术原理 ·········· 238
8.2 保护路径计算与使用 ·········· 240
8.2.1 实验环境 ·········· 241
8.2.2 LFA 路径技术 ·········· 242
8.2.3 Remote LFA（R-LFA）路径技术 ·········· 244
8.2.4 Extended R-LFA（ER-LFA）路径技术 ·········· 250
8.2.5 Topology Independent LFA（TI-LFA）路径技术 ·········· 253
8.2.6 TI-LFA 路径的数据面验证 ·········· 262
8.3 防微环路 ·········· 265

8.3.1 TI-LFA 路径与微环路 …………………………………………… 265
8.3.2 TI-LFA 路径与本地防微环路 …………………………………… 267
8.3.3 TI-LFA 路径与 SR 防微环路 …………………………………… 270
8.4 计算保护路径的仲裁策略 ………………………………………………… 284
8.4.1 实验环境 …………………………………………………………… 285
8.4.2 链路保护 …………………………………………………………… 286
8.4.3 节点保护 …………………………………………………………… 287
8.4.4 SRLG 保护 ………………………………………………………… 290
8.4.5 Node+SRLG 保护 ………………………………………………… 292

第1章

绪 论

1.1 引言

在打开本书的后续内容前,您可能会首先思考这样几个问题:

"我们为什么要关注多协议标签交换(Multi-Protocol Label Switching,MPLS)和分段路由(Segment Routing,SR)技术?"

"MPLS 和 Segment Routing 技术二者关系如何?"

我们首先来回答第一个问题。笔者之所以聚焦这两类技术撰稿成书,与您翻开这本书的目的一样,即运用这两项技术。诚然,MPLS 和 Segment Routing 都是具有明确需求指向的实用化技术,前者是为了解放传统 IP 网络转发性能,而后者则是简化 IP/MPLS 网络协议和解决基于 OpenFlow 的传统软件定义网络(Software Defined Network,SDN)的固有弊病。

关于第二个问题,既然我们将 MPLS 和 Segment Routing 两种技术放入同一本书的框架内,显然二者具有较强的内在联系,这种内在联系即 Segment Routing 技术脱胎于 MPLS、继承了其引入的标签机制,并将其与自身的源路由机制结合,实现了对 MPLS 网络的大规模精简和对骨干网 SDN 的革命性创新。

下面,我们从 MPLS 与 Segment Routing 技术的起源与发展讲起。

1.2 MPLS 与 Segment 技术起源与发展

1.2.1 MPLS 技术起源、发展与不足

为消除 IP 原有最长前缀匹配和逐跳查表导致的数据转发速率瓶颈，借鉴异步传输模式（Asynchronous Transfer Mode，ATM）信元交换原理，提出了 MPLS 技术。与传统 IP 转发方式相比，只在网络边缘分析 IP 报文头，而不用在每一跳都重复相同的操作，节约了处理时间。为此，MPLS 引入了转发等价类（Forwarding Equivalence Class，FEC）的概念，将具有相同转发处理方式的分组归为同一类，并为其赋予标签作为唯一标识，使得网络转发节点可通过匹配标签识别转发数据所归属的类，拥有相同标签即视为同类，并对其采取一致的转发处理；同时，因标签长度固定，通过简单的标签交换即可完成数据查表匹配转发，有效提升了转发速度。对网络转发有相同需求的上层协议使用相同的标签引导数据转发，使得 MPLS 技术可扩展支持 IPv6、网际报文交换（Internet Packet Exchange，IPX）、无连接网络协议（Connectionless Network Protocol，CLNP）等多种协议场景，这也即 MPLS 中的"MP"所标识的"Multi-Protocol"的来源。

然而，随着专用集成电路（Application Specific Integrated Circuit，ASIC）技术的发展，路由查找速度已非网络发展瓶颈，使得基于 MPLS 提升转发速度的需求不再旺盛。而且，支撑 MPLS 技术的背后，也存在若干不容忽视的问题，主要是其标签分发与运用过于复杂。MPLS 技术虽然引入了标签机制，但其本身只使用 MPLS 标签而不分配标签，通常场景下使用标签分发协议（Label Distribution Protocol，LDP）分发标签，在 MPLS 流量工程（Traffic Engineering，TE）场景下使用资源预留协议（Resource Reservation Protocol，RSVP）分发标签，而这两种标签分配协议都在事实上不必要地（经 Segment Routing 实践验证）增加了网络复杂度。

因此，许多场景下，MPLS 管理和部署被认为过于复杂。此外，为用户不同应用提供差异化网络服务的细粒度网络控制需求也逐渐凸显，而实施细粒度管控将进一步增加 IP/MPLS 网络复杂度。

1.2.2　Segment Routing 技术起源与发展

MPLS 虽存在 LDP/RSVP 标签分发配套组件资源消耗多、扩展性差的问题，但其本身引入的标签机制确实有益，为避免"因噎废食"，技术人员开始研究如何以新的简化标签分发协议取代 LDP/RSVP。与此同时，为改善传统 IP 网络的可扩展性、可管控性、安全性、移动性、服务质量，SDN 作为一种新型网络架构地位逐渐凸显，其通过在路由器、交换机等数据转发单元之上引入控制器，作为全网集中控制单元，实现了原集成于路由器的控制与转发两大网络子功能的解耦，为用户意图与服务质量的可编程集中优化提供了新型部署范式，正在为网络体系架构带来下一场革命。经典 SDN 采用了基于应用面、控制面和数据面的 3 层架构，其南向接口协议通常采用广为人知的 OpenFlow 协议，引入了"流"（Flow）的概念，即通过提取 MAC 地址、IP 地址等公共特征，将一次网络通信中产生的大量数据分组按特征抽象分类为不同的"流"，使网络设备无差别地看待同一条流的数据分组，实现了处理效率的提升。

但在 SDN 提升网络可编程性和灵活性的同时，也暴露了一些新问题。一是过于依赖控制器，其中心化部署逻辑决定了网络核心集中于控制器，控制器单点故障将无法承受。二是控制面和数据面交互过于频繁，在数据面每次发生网络故障时均需要控制器来恢复，一方面可靠性较差，另一方面控制器恢复时均须重新计算流的转发逻辑，并更新网络中大部分设备的流表，不利于网络规模扩展。三是性能提升存在瓶颈，数据传输依赖流表下发，而流表下发有速率瓶颈，尤其是 OpenFlow 控制粒度过小，每条流的每一跳转发都需要建立一条流表项，试想这样的场景：为在整个网络上建立起一条流的转发，需要对其转发路径上的所有路由器节点编程下发流表项；当流的规模上升到数百万时，需要的状态信息维持将难以实现。四是 SDN 颠覆了传统网络，设备厂商和运营商还未对其达成普遍共识，支持度和覆盖率有待提高。

Segment Routing 概念与技术体系，于 2013 年被首次提出，其起源于源路由，吸收了 MPLS 技术的标签机制和 SDN 技术的集中优化思路，通过"对网络协议做减法"删减了 MPLS 配套组件中为人诟病的 LDP 和 RSVP，又通过将数据面节点的分布式智能与控制面的集中式优化结合，一定程度上消解了网络对控制器的过度依赖，还将状态信息从整个网络中去除，使其仅存在于域入口节点（源节点）和数据包报头。换言之，Segment Routing 技术对 MPLS 和 SDN"取其精华、去其糟粕"，对网络协议、网络状态进行了全面删减，重构了 IP 网络体系，其超大规模组网能力使之可代替 MPLS，因此也被称为"下一代 MPLS"。

2015 年，Segment Routing 技术加速成熟，特别是在思科路由器正式支持 Segment Routing 流量工程后，催化了 SR 在不少早期用户网络中的部署，同时，越来越多的业界厂商因思科推动而开始对其宣布支持；同年，开源 SDN 控制器 Open Daylight 宣布支持该技术，是 Segment Routing 获得业界共识的一个标志性事件。

自提出后短短数年，Segment Routing 技术因其源路由、无状态、易部署、可跨域、快收敛等优良特性，充分体现了"应用驱动网络"的先进网络设计理念，已获得业界的广泛支持，2016 年已在国内外顶级运营商及 OTT（Over The Top）网络中开始实际部署。2017 年 2 月，Linux 内核 4.10 版本正式支持 Segment Routing 技术，这极大拓展了其应用场景，意味着基于主机，甚至容器就可以调度指定业务流在全网的端到端路径，为 Segment Routing 技术后续发展注入了强大动力。

1.3　Segment Routing 技术基础原理

在介绍 Segment Routing 技术基础原理之前，我们首先介绍几个作为原理根基的 Segment Routing 基本概念。

1.3.1　基本概念

1. Segment

直译为"段"，其本质是节点针对所收到数据包要执行的指令，如"将数据包通过特定接口转发""将数据包发送至指定服务实例"等。

2．SID（Segment Identifier）

Segment 的标识符。

3．Segment List

直译为"段列表"，也称"SID 列表"，是 Segment 的有序列表，其本质是中间节点对所收到数据包要执行的指令列表，可引导数据包通过网络的任意路径，该路径不受最短路径、域边界、路由协议等因素制约。

4．Segment 列表操作

SID 作为 SR 基础要素，其列表形式则表示了一个路径段组合。SR 将多个路径段组合在一起，构成了一条路径集合，从而实现了灵活的路径编排与管理。组合在一起的路径段由一系列 SID 组合标识，这个 SID 组合称为 SID 列表或段列表。携带 SID 列表的数据分组在转发过程中存在 3 种不同的处理模式。

- PUSH SID：按照预先定义的路径调整策略，在满足条件的数据分组中压入 SID 或者 SID 列表，并指定活跃 SID，进而指引后续节点按照活跃 SID 进行数据操作和转发。虽然数据分组携带多个 SID，但在某一时刻只有一个用于指导节点的数据转发，这个 SID 被称为活跃 SID。
- CONTINUE：如果当前活跃 SID 持续有效，则不对 SID 做任何操作，仅按照活跃 SID 对数据进行转发处理。
- NEXT SID：如果当前活跃 SID 使命完成，则将活动 SID 弹出，令 SID 列表中的下一个 SID 成为活跃 SID，并进行数据分组的操作转发。

5．全局 Segment

即在 SR 域（由参与 SR 模型的节点构成的网络域）内通用化、可识别的 Segment 标识，其中"全局"即指在 SR 域内。若 Segment A 为全局 Segment，则 SR 域内所有节点都知道如何对其转发处理。

6．本地 Segment

即仅在本地生效的 Segment，只有生成本地 Segment 的节点，在其转发表中安装与该本地 Segment 相关联的指令，并支持与该本地 Segment 相关联的指令。

1.3.2　Segment Routing 技术原理

　　Segment Routing 技术的本质是一种数据包交换技术,即由发起数据交换的路由器、主机或设备(均视为 SR 域节点)选择并生成路径,即指定流量途径的若干个关键节点,并基于段标识引导数据包沿该路径通过中间节点到达目的节点。Segment Routing 技术引导数据包的具体方式是在数据包报头位置插入带顺序的 Segment 列表,以指示收到这些数据包的中间节点如何按 Segment 列表操作处理和转发这些数据包。

　　可见,Segment Routing 技术的关键特征之一是源路由机制,由于指令被编码在数据包报头,而非下发至每个数据面交换设备,因此,转发路径不必像在基于 OpenFlow 的 SDN 那样,为所有可能经过的流维持状态信息,大幅降低了网络状态。这种网络简化丝毫不会影响网络控制性能,源节点通过在数据包报头添加适当 Segment 即可实现单条流颗粒度的数据包引导,在 IP/MPLS 网络中提供细粒度流量引导能力,在控制粒度和运营效率间找到了较好的平衡点,具备较强可扩展性,有力地支撑了 IP 网络端到端流量任意调度和网络软件定义可编程重构,成为基于 IP 网络的 SDN/NFV 下一代发展关键技术。

1.3.3　Segment Routing 架构

　　Segment Routing 从 SDN 架构孵化而来,其也采用了 SDN 将控制面和数据面解耦分离的设计思想。参考 SDN 的典型 3 层架构(应用面+控制面+数据面),由于特定的 Segment 应用不在我们研究范围之列,本书着重关注其控制面和数据面。需要注意的是,Segment Routing 架构主要基于已有的控制面和数据面拓展实现,而非对其进行了"全新定制"。

1. Segment Routing 控制面

　　目前,Segment Routing 控制面对内部网关协议(Interior Gateway Protocol,IGP)和边界网关协议(Border Gateway Protocol,BGP)均可支持,其中,可支持的 IGP 包括中间系统到中间系统(Intermediate System to Intermediate System,

IS-IS）和开放最短路径优先（Open Shortest Path First，OSPF）两类。由 IGP 分发的 Segment 称为"IGP Segment"，由 BGP 分发的 Segment 称为"BGP Segment"。值得注意的是，这些 Segment 可按需灵活组合，以满足用户特定应用需求。

（1）IGP Segment

主要分为两类，一类是用于指向特定前缀的 IGP 前缀 Segment，另一类是用于指向特定节点邻接链路（集合）的 IGP 邻接 Segment。

① IGP 前缀 Segment

也称"IGP Prefix Segment"，缩写为"Prefix Segment"或"PrefixSID"，与该 IGP 通告的一条前缀相关联，其代表的具体指令是"将收到的数据包沿着支持等价多路径路由（Equal Cost Multi-Path，ECMP）的最短路径，发往与该 IGP Prefix Segment 相关联的前缀"。IGP 前缀 Segment 是全局 Segment，具体可分为两类。一类是 IGP 节点 Segment，也称"Node Segment"或"NodeSID"，即与标识特定节点的前缀相关联的 Segment，通常为该节点环回地址的主机前缀；另一类是 IGP 任播 Segment，也称"Anycast Segment"或"AnycastSID"，即对多个节点分配相同的单播前缀，也即将该单播前缀转换为任播前缀，并将其视为一个"组前缀"，可为节点提供故障保护和负载均衡。

② IGP 邻接 Segment

也称"IGP Adjacency Segment"，缩写为"Adjacency Segment""Adjacency-SID"或"AdjSID"，即与单向邻接或单向邻接集合相关联的 Segment，其代表的具体指令是"将收到的数据包，由与该 Segment 相关联的邻接链路或邻接链路集合转发出去"。之所以设计这类 Segment，是为了将流量强制引导至指定链路，而无视 IGP 或其他规则规定的最短路径路由，通常可用 Adjacency Segment 引导流量经由直连链路去往邻居节点。该类型 Segment 虽可用作全局 Segment，但通常被用作本地 Segment，也可分为两类。一类是二层 Adjacency-SID，可为每个链路聚合组（Link Aggregation Group，LAG）成员链路分配二层 Adjacency-SID，以引导流量通过特定单条物理链路；另一类是组 Adjacency-SID，为路由器间的多条平行非 LAG 链路共同分配组 Adjacency-SID（Group Adjacency-SID），可使流量在组内链路上负载均衡。

（2）BGP Segment

BGP 相比 IGP 更简单，且覆盖率更广，因此被广泛使用。主要分为两类，一类是与 IGP 前缀 Segment 相对应的 BGP 前缀 Segment，另一类是用于指向 BGP 对等体的 BGP 对等体 Segment。

① BGP 前缀 Segment

也称"BGP Prefix Segment"或"BGP PrefixSID"，与 IGP 前缀 Segment 类似，BGP 前缀 Segment 与某个 BGP 前缀相关联，也是全局 Segment，可分为两类，一类是 BGP 非任播 Segment，其代表的具体指令是"将收到的数据包沿着支持 ECMP 的最优路径，发往与该 BGP Prefix Segment 相关联的前缀"，即将流量在可用的 BGP 多路径上负载均衡；另一类是 BGP 任播 Segment，使一组节点通告相同的 BGP 前缀，可提供类似 IGP 任播 Segment 的粗粒度流量引导能力。

② BGP 对等体 Segment

也称"BGP Peer Segment"，与 BGP 对等体会话（Peer Session）的特定邻居或邻居集合相关联，为本地 Segment，可分为 3 种。一是 BGP 对等体节点 Segment（BGP Peer Node Segment），与对等体会话的邻居相关联，其代表的具体指令是"将数据包经由 ECMP 多路径发往特定 BGP 对等体节点"；二是 BGP 对等体邻接 Segment（BGP Peer Adjacency Segment），与到特定邻居节点的特定链路相关联，其代表的具体指令是"将数据包经由到特定对等体节点的特定接口发往该对等体节点"；三是 BGP 对等体集合 Segment（BGP Peer Set Segment），与一组对等体会话的邻居相关联，其代表的具体指令是"将数据包经由 ECMP BGP 多路径发往属于特定对等体集合的 BGP 对等体节点"。

2．Segment Routing 数据面

Segment Routing 数据面可基于 MPLS，也可基于 IPv6，以下对两种数据面进行分别介绍。

（1）SR MPLS 数据面

Segment Routing 数据面基于 MPLS 时，可称为 SR MPLS 数据面，可直接利用现有 MPLS 架构，此时 SID 体现为 MPLS 标签栈或 MPLS 标签空间中的索引，Segment 列表在 MPLS 数据包中被表示为 MPLS 标签栈，应注意，SR MPLS 数据面对 IPv4 和 IPv6 均可适用，IPv4 和 IPv6 数据面都可对 MPLS 转发表项进行编程。

SR MPLS 标签栈操作按"Segment 列表操作"所述，可分为 PUSH、CONTINUE 和 NEXT 3 类，分别对应原 MPLS 数据面的 PUSH、SWAP 和 POP 3 种操作。

（2）SRv6 数据面

Segment Routing 数据面基于 IPv6 时，通常被称为 SRv6，此时通过定义分段路由头（Segment Routing Header，SRH）作为承载额外路由信息的扩展报头，Segment 体现为 IPv6 地址，Segment 列表被编码为 SRH 中含多个 IPv6 的有序列表。

1.4　Segment Routing 应用前景

放眼未来，随着 5G、云计算、边缘计算、物联网等新兴信息技术深刻驱动信息网络转型，Segment Routing 可面向 MPLS 网络升级、数据中心、大规模互联、云端融合、5G 网络等多种场景提供引领和支撑，具有无比广阔的应用前景。

1.4.1　Segment Routing 用于现有 MPLS 网络升级

Segment Routing 可基于对现有 MPLS 网络的控制面协议和数据面的拓展，实现增量部署、平滑演进，为原 MPLS 网络提供与上层应用的快速交互能力，利用集中式优化与分布式智能的优化组合，助力 SDN 打破其流表下达等性能瓶颈，重塑新型网络。

1.4.2　Segment Routing 用于数据中心

将 Segment Routing 的精准流量引导能力用于数据中心，可在负载均衡效率、网络路由不均衡性感知、性能路由和确定性网络探测等方面提供帮助，提升数据中心实施流量工程的能力。

1.4.3　Segment Routing 用于大规模互联

SR MPLS 可扩展网络，仅需数万条转发表条目，即可支持数十万个节点和数千万个底层物理端点实现大规模互联，而传统 MPLS 网络需要数百万条转发表条目。可将网络分为处于中央的核心域和处于边缘的叶子域，叶子域间通过核心域互联，若要进一步扩充网络，还可设置二级叶子域。

1.4.4　承载网络

Segment Routing 维护数据信息较少，可针对运营商 5G 网络中的差异化多重业务需求提供多种业务实例支持，还支持拓扑无关无环备份快速重路由（Topology-Independent Loop-Free Alternate Fast Reroute，TI-LFA FRR），具有较强的网络控制、网络隔离、快速恢复和泛在连接能力，固定宽带、固网、云、政企专线综合承载将随之增强，进一步促进云端融合。

1.4.5　SD-WAN 构建

软件定义广域网（Software Defined Wide Area Network，SD-WAN）实质是 SDN 面向广域部署的网络技术，要求部署方便、迭代快速，注定不能使用传统 WAN 中的 LDP 等复杂的分布式协议，而 Segment Routing 可为其提供简洁明了的自定义选路功能，且易于部署和维护，因此是 SD-WAN 构建的优质选项。

1.5　本书内容组织

Segment Routing 技术作为 IP 网络的重大技术创新，成为近年来少有的焦点项目，为网络提供强大的可编程能力，开启了网络体系创新的新时代。SR 技术不是单点技术，而是自成体系的网络架构，以 SR 架构为基础的特性拓展和应用层出不穷。Segment Routing 核心技术依托 MPLS 和 IPv6 作为数据面，通过拓展

OSPF、IS-IS、BGP 实现段信息分发，其路径编排能力可以独立部署，也可以通过 ODL 等集中式控制器进行编排。本书以 Segment Routing 技术实验为指导，结合 Segment Routing 技术原理讲解，通过搭建实验场景，对 Segment Routing 技术进行配置、调试、协议抓包分析等实验，提高相关人员对协议机制的认知深度，从而快速有效地掌握 Segment Routing 技术原理、配置操作以及部署应用。

　　本书以 Segment Routing 技术为主要内容，对 MPLS 基本功能和基于 MPLS 的 Segment Routing 技术提供实验指导。全书分为 3 个部分，第 2、3、4 章为基础实验部分，阐述基于 EVE-NG 的实验环境构建、MPLS 基础技术实验以及基本的 LDP 技术实验；第 5、6、7 章为路由协议的 Segment Routing 功能拓展，重点围绕 OSPF、IS-IS 和 BGP 的 Segment Routing 功能开展实验；第 8 章主要讲解基于 Segment Routing 的快速重路由技术，通过对比传统 LFA 路径和 TI-LFA 路径技术，使读者掌握 Segment Routing 技术在快速重路由领域带来的特色优势。

第 2 章

基于 EVE-NG 的实验环境构建

2.1 EVE-NG 概述

EVE-NG 是当前优秀的多厂商网络仿真软件之一，受到网络工程和信息安全等专业人士的喜爱，其图标如图 2.1 所示。通过它，可以在接近真实的实验环境中模拟各类设备组成的大型网络，验证所学到的知识，设计个性化的网络架构。

图 2.1 EVE-NG 图标

EVE-NG 能够同时支持 Dynamips、IOL（IOS on Linux）格式的思科网络设备镜像和 QEMU 格式的通用镜像，可以围绕虚拟设备使用并将其与其他虚拟或物理设备互连，提升了网络仿真实验的可用性、可重用性、可管理性、互联性，为网络工程师等专业人员提供不依赖于物理设备的学习和实践手段。EVE-NG 官方提供 ISO 安装镜像和开放虚拟化设备（Open Virtualization Appliance，OVA）镜像，用户可以直接安装在物理机、裸金属服务器上或导入虚拟机管理软件中运行。EVE-NG 基于 Web 界面进行虚拟网络拓扑的搭建，只要有相关厂商的设备镜像，就可以使用 EVE-NG 构建如图 2.2 一样大型的，具备多个厂商、多个虚拟和物理设备的拓扑，进行学习实验或技术验证。

第 2 章 基于 EVE-NG 的实验环境构建

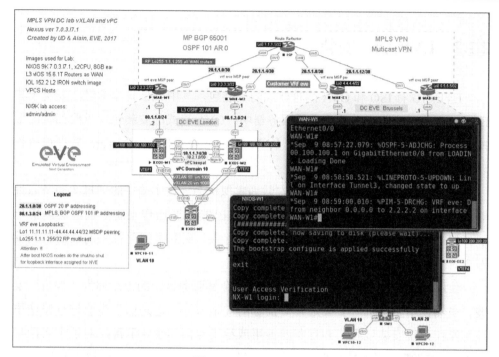

图 2.2　EVE-NG 示例

2.1.1　EVE–NG 的用途

EVE-NG 不仅可以模拟思科（Cisco）、新华三（H3C）、华为、瞻博（Juniper）等厂商的网络设备，也可以运行虚拟机。理论上来说，只要能将想要运行的虚拟设备磁盘转换为 EVE-NG 支持的格式，就可以在 EVE-NG 中运行。借助 EVE-NG 多样化的镜像支持能力和强大的兼容性，可以快速构建包含多种品牌和设备类型的拓扑架构，解决经费有限导致实验无法进行的问题，免去了以往只能使用物理设备搭建实验环境而进行的布线等冗繁步骤。在生产环境的网络搭建、改造和运维中，可以使用 EVE-NG 对设计思路和方案进行测试验证和更改，简洁迅速地在安全环境中重现和改进真实架构，而无须触及真实网络，降低故障风险，并且在建设过程中快速导入虚拟设备中的配置，提高工作效率，同时，还可以通过重新创建问题网络定位和排除故障。

2.1.2 EVE-NG 安装

EVE-NG 有 ISO 和 OVA 两种格式的安装包,其中 ISO 光盘镜像主要用于将 EVE-NG 安装到物理机或裸金属服务器中;OVA 格式的安装包将需要部署的虚拟机的磁盘和配置等文件封装成一个单一的包,可直接在 VMware 等虚拟机管理软件中部署 VM。

使用 OVA 安装包将 EVE-NG 安装为虚拟机会导致实验中的虚拟实验节点性能的下降,因为实验部署的所有节点都在 EVE-NG 操作系统中被嵌套虚拟化了。但是只要宿主机满足部署节点的资源需求,嵌套虚拟化带来的负面影响就不会对实验造成拖累。

使用 ISO 镜像直接安装在物理机或裸金属服务器中的方法称为"裸机"安装。虽然使用"裸机"安装是官方推荐的安装方法,但是由于需要进行镜像引导等操作系统安装过程,并且需要占用或租用专门的硬件资源,对于个人用户来说成本较高,便携性较低,所以本书只对 OVA 安装包在虚拟机管理软件 VMware Workstation 上的安装过程进行介绍。

在进行安装之前,我们需要了解 EVE-NG 支持的虚拟化平台及软件。

- VMware Workstation 12.5 或更高版本
- VMware Player 12.5 或更高版本

官方推荐的笔记本计算机/台式计算机系统要求支持虚拟化并已经在 BIOS 开启 Intel VT-x/EPT,操作系统采用 Windows 7、Windows 8、Windows 10 或 Linux 桌面版,推荐配置见表 2.1。

表 2.1 EVE-NG 推荐配置

宿主机硬件要求		EVE-NG 虚拟机资源分配
处理器	Intel i7(8 个逻辑处理器),支持并开启 Intel 虚拟化	8/1(处理器数量/每个处理器的核心数量),并且启用 Intel VT-x/EPT 虚拟化引擎
内存	32GB	24GB 以上
硬盘空间	200GB	200GB 以上
网络要求	局域网或广域网	VMware NAT 或 Bridged 网络

笔记本计算机/台式计算机能够运行中小型实验。虚拟节点的性能和数量取决于实验中部署的节点类型。

IOL 镜像：每个实验最多 120 个节点。

vIOS 镜像：每个实验最多 40 个节点。

CSR 镜像：每个实验最多 10 个节点。

其中，IOL 是思科的 IOS（Internetwork Operating System）on Linux 的缩写，它是 IOS 在通用 x86 平台上进行编译生成的网络设备镜像，在思科内部用于网络设备模拟；vIOS 是 IOS 的虚拟化版本，它可以作为一个完整的虚拟机运行，具备完整的 3 层控制面和数据面功能；CSR 是思科云服务路由器（Cisco's Cloud Services Router）的虚拟镜像，可用于公有云和私有云的网络环境模拟。

在 VMware Workstation 上部署之前，需要从 EVE-NG 官网将 OVA 镜像下载至本地。打开 VMware Workstation，在主界面中单击"打开虚拟机"，然后在弹出的资源管理器界面中选中 OVA 镜像，如图 2.3 所示。

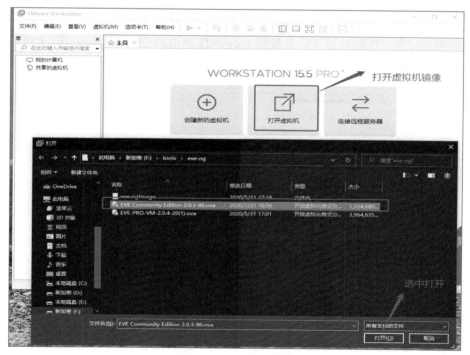

图 2.3　OVA 镜像导入

修改新虚拟机的名称并选择存储路径，单击"导入"，如图 2.4 所示。

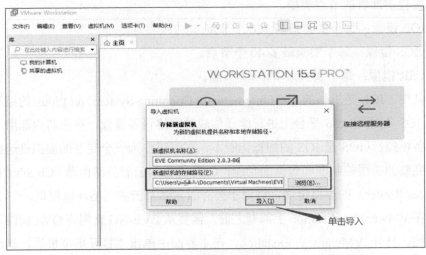

图 2.4　虚拟机存储路径

虚拟机导入完成后，在侧边栏选择修改虚拟机的配置，建议在保证宿主机性能和运行稳定性的前提下，尽量给 EVE-NG 虚拟机分配更多的内存、CPU 数量和硬盘空间，如图 2.5 所示。

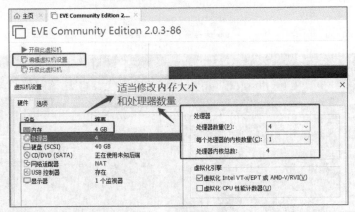

图 2.5　虚拟机资源配置

如果宿主机的网络接口有限，或者局域网内没有动态主机配置协议（Dynamic Host Configuration Protocol，DHCP）服务器，建议将网络连接模式选择为 NAT 模式，如图 2.6 所示。

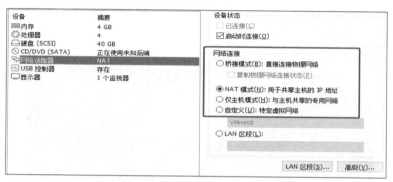

图 2.6　虚拟机网络模式修改

资源和网络配置完成后，开启虚拟机。开机过程中可以看到 EVE-NG 的启动界面，如图 2.7 所示。

图 2.7　EVE-NG 启动界面

这里需要注意的是，选择 NAT 模式或者桥接模式，如果 IP 地址不能正确分配到网卡，就会在开机界面卡住，这个时候按 F11 进入命令行开机界面，使用"Ctrl+C"的组合键中断地址分配，然后重启即可恢复正常。

虚拟机第一次启动，会有配置脚本提示进行初始配置，请自行根据实际情况进行配置。虚拟机正常启动后，在命令行前面几行显示默认 root 用户名、密码和访问地址，如图 2.8 所示。

图 2.8　EVE-NG 虚拟机用户名、密码和访问地址

请记住上面显示的 IP 地址，网络节点镜像上传、Web 管理和虚拟实验拓扑的搭建也是通过这个地址实现的。在浏览器（推荐 Chrome 或 Firefox）输入上述地址即可进入 Web 管理界面，如图 2.9 所示。

图 2.9　EVE-NG 的 Web 管理界面

至此，EVE-NG 在 VMware Workstation 上的部署已经完成，即将开始进行虚拟机仿真实验拓扑的搭建。

2.2　实验仿真环境构建

2.2.1　客户端集成软件包的安装

EVE-NG 自带 HTML5 控制台，无须额外安装终端软件就能进行 EVE-NG 中的虚拟设备配置；同时，EVE-NG 也支持使用本地安装的终端软件管理实验中的虚拟设备，使用数据包捕获等功能进行网络协议分析工作，安装 EVE-NG 官方提供的客户端集成包即可达到以上目的。

在 EVE-NG 官方平台下载 Window 平台客户端集成包 EVE-NG-Win-Client-Pack-2.0.exe，以管理员身份运行。注意安装时的提示，如果需要一并安装数据包捕获软件 Wireshark 和远程控制软件 UltraVNC，建议不要更改默认安装路径，如图 2.10 所示。

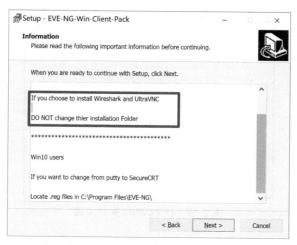

图 2.10　EVE-NG 客户端集成包安装信息

在安装组件选择界面，EVE-NG 已对必需的软件和脚本进行选择，保留了可选的 UltraVNC 和 Wireshark。建议将 UltraVNC 选项保留为选中状态，它是 EVE-NG 首选的 Windows VNC 客户端；如果系统中已完成 Wireshark 的安装并且版本较新，可以不选择安装 Wireshark，如图 2.11 所示。

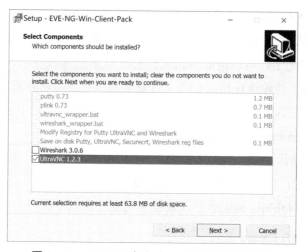

图 2.11　EVE-NG 客户端集成包安装组件选择

继续下一步，当要求选择 UltraVNC 选项时，用到的组件只有 UltraVNC Viewer 一个，如图 2.12 所示。

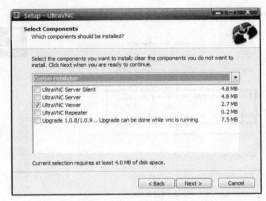

图 2.12　UltraVNC 安装步骤

所有的软件和脚本安装完毕后，会回到软件包的安装界面，最后分别选择"Next"和"Finish"，完成整个软件包的安装。

2.2.2　虚拟节点镜像管理

1. 导入网络节点镜像

EVE-NG 支持 IOL、Dynamips、QEMU 3 种类型的镜像，在虚拟实验节点实例化之前必须先将网络节点镜像上传至 EVE-NG 虚拟机，由于 EVE-NG 运行在 Ubuntu 系统之上，可以借助 WinSCP 进行 SCP 传输或使用 SecureFX 等工具进行 SFTP 传输，传输地址使用 EVE-NG 安装完成后虚拟机界面显示的地址，如图 2.8 所示。将 EVE-NG 支持的不同种类镜像分别上传至以下位置。

IOL（IOS on Linux）：/opt/unetlab/addons/iol/bin/。

Dynamips 镜像：/opt/unetlab/addons/dynamips/。

QEMU 镜像：/opt/unetlab/addons/qemu/。

IOL 镜像和 QEMU 镜像的导入过程中需要注意以下两点。

• IOL 镜像的 license 文件

IOL 仅在 Cisco 内部使用，所以 Cisco 官方不提供镜像下载，因此，运行

第 2 章 基于 EVE-NG 的实验环境构建

IOL 镜像不仅需要后缀为 bin 的镜像文件，还需要名为 iourc 的 license 文件。iourc 文件可以通过 CiscoIOUKeygen.py 的工具自动生成，所以在上传 bin 镜像文件的时候需要将 CiscoIOUKeygen.py 工具一并上传至 EVE-NG 的 IOL 存放目录，然后在 EVE-NG 终端中运行"python CiscoIOUKeygen.py | grep-A 1 'license' > iourc"命令生成 iourc 文件。

```
root@eve-ng:/opt/unetlab/addons/iol/bin#           //进入 IOL 镜像存放目录
root@eve-ng:/opt/unetlab/addons/iol/bin# python CiscoIOUKeygen.py |
grep -A 1 'license' > iourc                        //生成 iourc 文件
```

- QEMU 镜像的目录名称要求

QEMU 镜像是以 qcow2 为后缀的文件，文件名一般是 hda.qcow2 或者 virtioa.qcow2，这就要求存放镜像的目录名称要遵循一定的规则，请参照 EVE-NG 官方网站给出的目录命名规则。本书以实验需要用到的 XRv9k 虚拟路由器为例，先在/opt/unetlab/addons/qemu/目录下依据官方目录命名规则建立 xrv 开头的目录，后面可以在"-"之后，添加任何字符作为镜像的标记，如 xrv-9k-6.3.1。

```
root@eve-ng:/opt/unetlab/addons/qemu# mkdir xrv-9k-6.3.1
```

然后将 xrv-9k-6.3.1 的镜像上传至新建目录，如图 2.13 所示。

图 2.13　QEMU 目录示例

将镜像文件上传至 EVE-NG 后，还需要进行镜像权限的修正才能正常实例化虚拟实验节点，运行命令"/opt/unetlab/wrappers/unl_wrapper -a fixpermissions"修正权限。

```
root@eve-ng:/# /opt/unetlab/wrappers/unl_wrapper -a fixpermissions
```

2. Cisco XRv 虚拟路由器介绍

本书实验使用了思科 XRv9k 虚拟路由器，XRv9k 是思科支持分段路由特性较为全面的路由器系列，伴随着 SR 技术的研发，出现了多个迭代版本，如 5.2.2、6.0.1、6.3.1、7.1.1 等，版本号越大支撑的 SR 特性越丰富。

例如，在 6.0.1 版本中不存在 segment-routing local-block 的配置，也没有 ping/traceroute sr-mpls 功能，只能使用传统 mpls ping/traceroute 命令等。到了 6.3.1 版本，出现了 ping sr-mpls 命令。7.1.1 版本特性更为丰富，尤其是在支持路径计算单元通信协议（Path Computation Element Communication Protocol，PCEP）与 ODL 对接层面，但是耗费资源更多，一般 PC 在支撑多个虚拟实验节点同时运行时比较吃力。因此，在不做特殊说明的情况下，本书默认使用 6.3.1 版本进行实验。

2.2.3　Lab 管理

1. 创建 Lab

完成实验节点镜像的导入后，就可以在浏览器中通过 EVE-NG 的图形化界面进行实验拓扑的生成和网络的配置。EVE-NG 启动后，用浏览器打开 EVE-NG 的 Web 首页，登录 http://192.168.81.130，显示如图 2.9 所示的登录界面，输入默认的用户名 admin，密码 eve，由于前面已经完成客户端集成软件包的安装，因此这里的登录方式选择 Native console，通过本地软件进行 EVE-NG 实验中虚拟实验节点的配置和管理。

页面登录完成会进入 EVE-NG 的主管理界面，页面上方是系统管理菜单栏，通过相应的按钮可以实现主页导航、系统（用户）管理、系统状态显示和管理以及官方帮助的导航。在主管理界面中间占据主要显示位置的是文件管理区域，实验过程中建立的每个场景、设计的每个拓扑，在 EVE-NG 中都被定义成一个 Lab，文件管理区域就是对 Lab 进行创建和管理的地方，可以在文件管理区域单击相应的按钮进行 Lab 的创建、重命名、移动和删除，以及 Lab 的导入和导出，如图 2.14 所示。

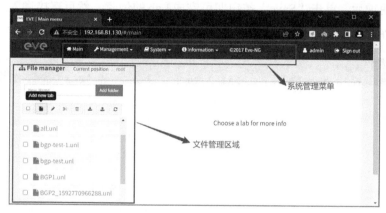

图 2.14　EVE-NG 主管理界面

按钮下方是已经建好的 Lab，单击相应的名字，然后在右边区域选择"Open"即可载入已有 Lab 进行实验的配置，如图 2.15 所示。

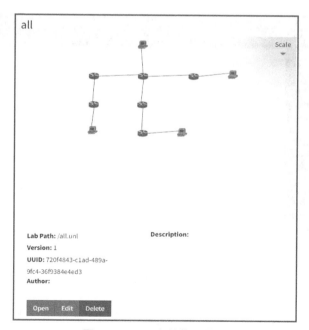

图 2.15　Lab 文件载入界面

下面新建一个名为 new 的新 Lab 进行 Lab 操作界面的介绍。在文件管理区域单击"Add New Lab"按钮，填入 Lab 名称和版本号两个必填内容，以及一些附加说明项后，单击"Save"按钮即可完成 Lab 的创建，然后直接进入这个 Lab 的操作界面，如图 2.16 和图 2.17 所示。

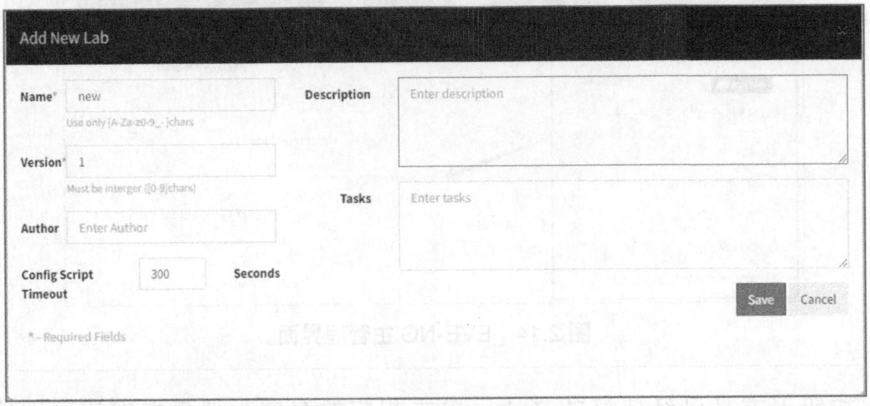

图 2.16　新建 Lab 界面

图 2.17　Lab 操作界面

在操作界面左侧单击"Add an object"按钮，添加虚拟实验节点、网络对象、导入图片、自定义图形和文本，在右侧画布区域单击鼠标右键也可以弹出

"Add a new object"菜单，完成上述各类对象的快捷添加，如图2.18所示。

2．生成实验拓扑

要生成实验拓扑，需要先添加虚拟实验节点，通过单击"Add an object"的"Node"菜单，在实验中添加交换机、路由器、虚拟PC以及虚拟机等多种EVE-NG支持的虚拟实验节点。在弹出的"ADD A NEW NODE"菜单"Template"栏中，显示所有EVE-NG支持的虚拟节点类型，需要使用某一

图2.18　添加对象菜单

类的虚拟节点，必须要先导入相应的节点镜像，没有导入的设备在列表中是灰色不可选的状态；已导入的设备显示为蓝色，可以被正常添加。可以看到之前导入的Cisco XRv类型已是可选状态，如图2.19所示。

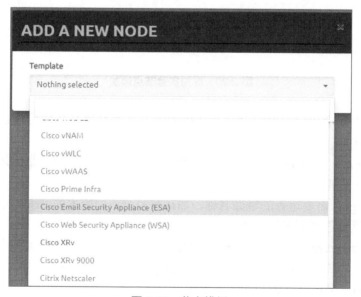

图2.19　节点模板

单击"Cisco XRv"选项进行xrv9k虚拟实验节点的添加，如图2.20所示。

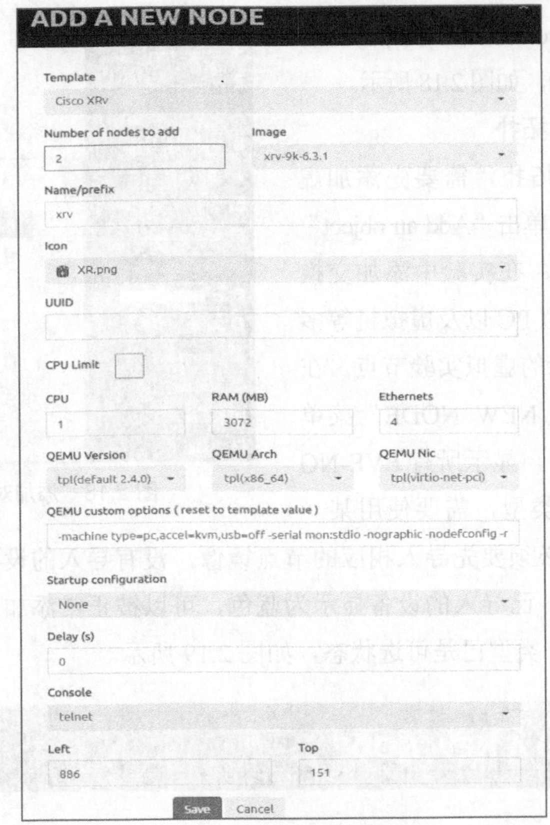

图 2.20　虚拟实验节点添加

下面的几个选项需要根据运行环境的情况进行设置，其余选项建议保持界面中的默认值。

- Number of nodes to add：添加的节点数量，可以一次批量添加多个同类型的虚拟实验节点。
- Image：需要运行的镜像名称，如果有多个不同版本的 Cisco XRv 镜像，这里要选择正确的名称。
- Name/prefix：节点在 Lab 中的名字或前缀，如果一次添加多个虚拟实验节点，EVE-NG 会在这个值后面自动添加数字组成节点名字。
- Icon：选择节点在 Lab 中的图标。
- CPU Limit：选择此项可以限制 EVE-NG 进程的 CPU 使用率，释放处理器资源，如果 Lab 中运行的虚拟实验节点较多，建议开启此选项，供计算机上可能运行的其他任务使用 CPU。

第 2 章　基于 EVE-NG 的实验环境构建

- CPU：设置虚拟实验节点的 CPU 数量，建议保持默认。
- RAM(MB)：设置虚拟实验节点的内存，建议保持默认。
- Ethernets：虚拟实验节点的接口数量，根据需要进行修改。
- Startup configuration：虚拟实验节点的启动配置文件，None 表示空配置，Exported 表示从已导出的配置启动，如图 2.21 所示。

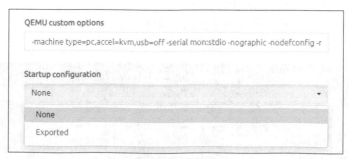

图 2.21　启动配置

- Console：连接虚拟实验节点使用的协议，有 telnet、vnc、rdp 3 种类型，依据实际情况进行选择。
- Left 和 Top：设置虚拟实验节点在画布区域的坐标，由于 EVE-NG Lab 中的图标可以自由拖动，这两项可以保持默认。

上述节点配置完成后，单击"Save"生成虚拟实验节点。在生成的节点图标上面单击鼠标右键，可以进行节点启动、停止、清除配置、导出配置、捕获数据包、编辑节点配置和删除节点等操作，如图 2.22 所示。

单击"Start"按钮将 xrv2 启动，图标将变为蓝色并且下方的运行状态显示将由停止转为运行。除了使用单击图标进行单个虚拟实验节点的操作，EVE-NG 还支持框选多个虚拟实验节点设备进行批量操作的功能。

图 2.22　节点配置选项

实验过程中对虚拟实验节点的配置是通过控制台连接命令行，然后与传统物理设备一样输入命令实现的。虚拟实验节点启动图标变为蓝色后，左键单击节点图标，由于在登录 EVE-NG Web 界面时选择 Native console 方式并且节点生成时配置 Telnet 控制台连接，因此在浏览器（以 Chrome 为例）窗口中会弹出连接提示，选择"打开 SSH，Telnet and Rlogin client"，进入本地控制台，如图 2.23 所示。

图 2.23　本地控制台弹出提示

控制台中会打印虚拟实验节点的初始化和版本等信息，待初始化完成后，即可在控制台中输入命令对节点进行配置，如图 2.24 所示。

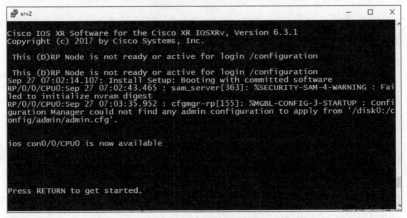

图 2.24　控制台界面

虚拟实验节点添加完成后，还需要将各节点按照实验拓扑连接起来，需要特别注意的是，在 EVE-NG 的社区版中需要将相应的节点关闭才能创建连接。鼠标指针移至关闭后的 xrv1 上，xrv1 的图标中会显示橙色插头图样，在插头上按住鼠标左键，拖曳连线至 xrv2 上，此时 xrv2 上会显示蓝色插头图样，然后释放左键，如图 2.25 所示。

图 2.25　连接设备

释放鼠标左键后，将弹出连接选项对话框，这里需要选择虚拟实验节点用哪个接口跟另外一个节点的接口建立连接。选择 xrv1 的 Gi0/0/0/0 接口与 xrv2 的 Gi0/0/0/0 接口连接，如图 2.26 所示。

第 2 章 基于 EVE-NG 的实验环境构建

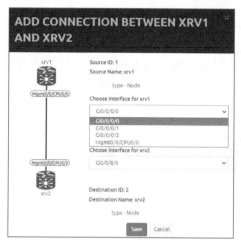

图 2.26 节点连接选项

成功建立连接后，在虚拟实验节点的图标上会显示连接使用的接口，如图 2.27 所示。至此，一个简单的实验拓扑建立完毕。

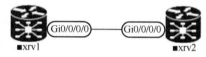

图 2.27 建立连接

3. 添加网络

EVE-NG 中支持添加"Network"，从而实现实验中多个节点的连接，以及 EVE-NG 内部虚拟实验节点与外部真实物理网络或设备的连接。在操作界面左侧单击"Add an object"按钮或在画布区域单击鼠标右键，选择"Network"，在弹出的对话框中进行网络类型的选择和设置，如图 2.28 所示。

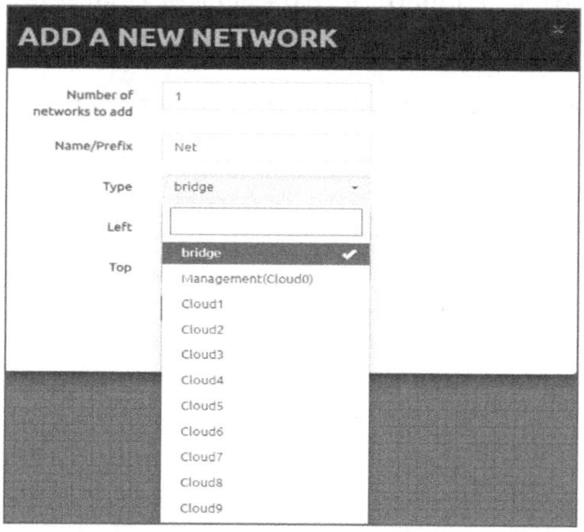

图 2.28 建立 Network

对话框中的"Number of networks to add""Name/Prefix""Left"和"Top"的含义与前文中虚拟实验节点添加时相同,"Type"选项主要有以下两种。

- bridge:bridge 网络就像一台非网管交换机,它支持传输带 vlan tag 的 dot1q 数据包,主要用于连接多个虚拟实验节点,实现多个节点间的通信,如图 2.29 所示。

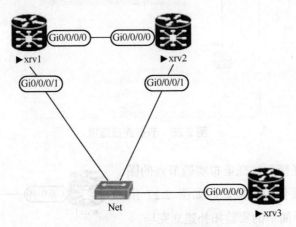

图 2.29 bridge 网络类型示意

- Cloud0~Cloud9:表示要桥接到 EVE-NG 服务器中的哪个以太网接口(本书运行环境为 VMware Workstation,因此 Cloud 类型网络桥接的是 VMware 的虚拟网卡),Cloud 类型网络是 EVE-NG 连接外部设备和网络从而实现虚实结合的关键。Cloud0~Cloud9 分别代表桥接至第 1 个至第 10 个以太网接口,Cloud0 是 EVE-NG 默认的管理接口,其对应关系在 EVE-NG 系统的/etc/network/interfaces 文件中定义。

```
root@eve-ng:~# cat /etc/network/interfaces
# This file describes the network interfaces available on your system
# and how to activate them. For more information, see interfaces(5).

# The loopback network interface
auto lo
iface lo inet loopback

# The primary network interface
iface eth0 inet manual
```

```
auto pnet0
iface pnet0 inet dhcp
    bridge_ports eth0
    bridge_stp off

# Cloud devices
iface eth1 inet manual
auto pnet1
iface pnet1 inet manual
    bridge_ports eth1
    bridge_stp off
```

"Network"添加完成后,将虚拟实验节点与其连接,可实现虚拟实验节点间的内部连接、虚拟实验节点与实验 Lab 外部设备的连接。

4. 数据包捕获

在实验的过程中,可以通过捕获虚拟实验节点上的数据包进行协议分析和配置验证,EVE-NG 支持调用 Wireshark 进行虚拟接口的数据捕获。

在运行的虚拟实验节点图标上单击鼠标右键,单击"Capture",选择需要进行捕获的接口,如图 2.30 所示。

第一次捕获数据包会要求缓存 EVE-NG 服务器的密钥,此时需要在 windows 命令提示符窗口输入"y"同意缓存,Wireshark 就会对流经接口的数据包进行捕获。

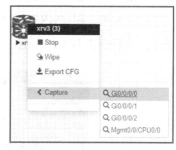

图 2.30 接口数据包捕获

需要注意的是,如果在 EVE-NG 系统初始化时修改了系统用户名和密码,Wireshark 启动之前会在 windows 命令提示符窗口提示"Access denied",这是因为客户端集成软件包的批处理文件中定义的用户名和密码与修改后的不同。进入软件包的安装目录(默认为 C:\Program Files\EVE-NG),对 wireshark_wrapper.bat 文件进行编辑,修改第 2 行和第 3 行为正确的用户名和密码,如图 2.31 所示。

观察 wireshark_wrapper.bat 文件的最后一行,可以发现调用的 Wireshark 绝对路径也在此定义,如果在安装客户端集成软件包的步骤中,没有选择安装

Wireshark 或者安装目录不在默认路径，需要在此文件中将 Wireshark 修改为正确的路径。

```
wireshark_wrapper.bat - 记事本
文件(F) 编辑(E) 格式(O) 查看(V) 帮助(H)
@ECHO OFF
SET USERNAME="root"
SET PASSWORD="eve"

SET S=%1
SET S=%S:capture://=%
FOR /f "tokens=1,2 delims=/ " %%a IN ("%S%") DO SET HOST=%%a&SET INT=%
%b
IF "%INT%" == "pnet0" SET FILTER=" not port 22"

ECHO "Connecting to %USERNAME%@%HOST%..."

"C:\Program Files\EVE-NG\plink.exe" -ssh -batch -pw %PASSWORD%
%USERNAME%@%HOST% "tcpdump -U -i %INT% -s 0 -w -%FILTER%" | "C:
\Program Files\Wireshark\Wireshark.exe" -k -i -
```

图 2.31 wireshark_wrapper.bat 文件

如果在第一次捕获数据包时没有缓存 EVE-NG 服务器的密钥，Wireshark 启动之前会弹出“connection abandoned”的错误信息。在 Windows 命令提示符窗口中进入软件包的安装目录（默认为 C:\Program Files\EVE-NG），输入“plink.exe -ssh -l USERNAME -pw PASSWORD HOST”进行缓存，其中 USERNAME 和 PASSWORD 是 EVE-NG 系统的用户名和密码，HOST 是 EVE-NG 的 IP 地址，如图 2.32 所示。

```
C:\Program Files\EVE-NG>plink.exe -ssh -l root -pw eve 192.168.81.130
Using username "root".
Access granted. Press Return to begin session.
Welcome to Ubuntu 16.04.3 LTS (GNU/Linux 4.9.40-eve-ng-ukms-2+ x86_64)

 * Documentation:
 * Management:
 * Support:
Last login: Tue Sep 27 16:37:15 2022 from 192.168.81.1
0;root@eve-ng: ~root@eve-ng:~#
```

图 2.32 使用 plink 进行 ssh 连接和密钥缓存

5. Lab 导出

EVE-NG 支持将创建好的 Lab 以及 Lab 中的节点配置一起导出。首先在要导出的 Lab 中，将各个虚拟节点的配置在相应终端中保存，保持虚拟节点是 Start 状态。在 Web 界面左侧选择“More actions”，单击“Export all CFGs”，关注浏览器页面右上角 Notifications 处，出现“Export All: done”即导出配置完成，如图 2.33 所示。

第 2 章 基于 EVE-NG 的实验环境构建

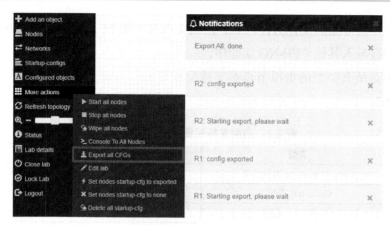

图 2.33 导出配置文件

此时，查看 Lab 页面左侧"STARTUP-CONFIGS"栏，可以看到每个虚拟节点的启动配置按钮由灰色变为 OFF，如果有导出配置不成功的，这里的开关为灰色不可选状态，如图 2.34 所示。

图 2.34 启动文件可用

导出配置完成后，关闭所有虚拟节点，关闭 Lab。返回主界面选择要导出的 Lab，在列表上方单击"Export"按钮，如图 2.35 所示。

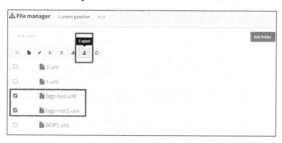

图 2.35 导出 Lab

最后，将导出的文件另存为 ZIP 格式文件到本地 PC，导出后的 ZIP 格式文件可以直接导入其他 EVE-NG 实验中。

并不是所有类型的虚拟节点都支持配置导出，当前支持配置导出的节点类型见表 2.2。

表 2.2 当前支持配置导出的节点类型

序号	类型	序号	类型
1	Cisco Dynamips 类型镜像	12	Juniper VRR 系列
2	Cisco IOL 格式	13	Juniper VMX 系列
3	Cisco ASA 系列	14	Juniper vMX-NG 系列
4	Cisco ASAv 系列	15	Juniper vQFX 系列
5	Cisco CSR1000v 系列	16	Juniper vSRX 系列
6	Cisco Nexus 9K 系列	17	Juniper vSRX-NG 系列
7	Cisco Nexus Titanium 系列	18	Mikrotik 系列
8	Cisco vIOS L3 系列	19	PFsense FW 系列
9	Cisco vIOS L2 系列	20	Timos Alcatel 系列
10	Cisco XRv 系列	21	vEOS Arista 系列
11	Cisco XRv9k 系列		

2.3 路由器配置管理

2.3.1 命令行工作模式

本书的实验场景采用思科 XRv9k 虚拟路由器作为虚拟实验节点，思科设备的命令行定义了 4 种工作模式。

- 一般用户模式：用户可以查看路由器的连接状态，访问其他网络和主机，但不能查看和更改路由器的设置内容。提示符为"Router >"。
- 特权模式：可以执行所有的用户命令，还可以看到和更改路由器的设置内容，不能对接口、路由协议进行配置。提示符为"Router# "，进入方式为在用户模式下输入命令"enable"。

```
Router>enable                          //进入特权模式
Router#
```

- 全局配置模式：主要用于配置路由器的全局性参数。提示符为"Router(config)#"，进入方式为在特权模式下输入命令"configure terminal"。

```
Router#configure terminal
Enter configuration commands, one per line.  End with CNTL/Z.
Router(config)#
```

- 全局配置模式下的子模式：如接口模式、路由引擎模式等，接口模式下可以进行 IP 地址配置、速率限制等接口设置，路由引擎模式下可以配置路由协议的相关参数。

在任意模式中，输入"exit"命令返回上一级模式，输入"end"命令则直接返回特权模式。需要注意的是，EVE-NG 中思科 XRv9k 虚拟路由器系统启动完成后会自动进入特权模式。

2.3.2 配置与保存

Cisco XRv9k 中大部分的配置命令和编辑与传统的 Cisco IOS 相同，但是在配置的管理方面 XRv9k 有着更强大的功能。区别于传统的 Cisco IOS 配置命令输入即生效的执行方式，XRv9k 中配置命令输入完成后，需要执行"commit"命令才能生效。生效之前的配置会在暂存区，需要经过有效性验证并取得授权，在未生效前都可以进行检查和修改，检查完成后输入"commit"将配置提交并生效。

```
RP/0/0/CPU0:ios#configure terminal
Sat Nov  5 12:42:14.907 UTC
RP/0/0/CPU0:ios(config)#hostname xrv1
RP/0/0/CPU0:ios(config)#show running-config
Building configuration...
!! IOS XR Configuration 6.3.1
!! No configuration change since last restart
!
interface MgmtEth0/0/CPU0/0
 shutdown
!
```

```
interface GigabitEthernet0/0/0/0
 shutdown
!
interface GigabitEthernet0/0/0/1
 shutdown
!
interface GigabitEthernet0/0/0/2
 shutdown
!
end

RP/0/0/CPU0:ios(config)#
```

上述命令尝试修改虚拟实验节点的 hostname 为 xrv1，配置完成后查看当前配置文件发现并未生效，命令行的提示符仍是"ios"。在输入"commit"使配置生效前可以对已输入的命令进行修改，在全局配置模式下输入"show configuration"查看哪些命令还未生效，如下。

```
RP/0/0/CPU0:ios(config)# show configuration
Building configuration...
!! IOS XR Configuration 6.3.1
hostname xrv1
end

RP/0/0/CPU0:ios(config)#
```

"show configuration commit list"命令可以查看每次"commit"生效的 ID、用户名、时间等信息，注意此时虚拟实验节点已经修改为 xrv1，表明配置已经提交生效，如下。

```
RP/0/0/CPU0: xrv1#show configuration commit list
SNo.  Label/ID     User   Line        Client   Time Stamp
~~~~  ~~~~~~~~     ~~~~   ~~~~        ~~~~     ~~~~~~~~~~
1     1000000003   eve    con0_0_CPU0 CLI      Sat Nov 5 12:48:34 2022
2     1000000002   eve    con0_0_CPU0 CLI      Sat Nov 5 12:48:02 2022
3     1000000001   eve    con0_0_CPU  CLI      Sat Nov 5 12:47:11 2022
RP/0/0/CPU0: xrv1 (config)#
```

使用"show configuration commit changes？"命令加上 Commit ID（即上面命令打印输出的"Label/ID"值），每次配置的更改内容也可查看。

```
RP/0/0/CPU0: xrv1#show configuration commit changes ?
  all          Display configurations committed for all commits
  last         Changes made in the most recent <n> commits
  since        Changes made since (and including) a specific commit
  1000000003   Commit ID
  1000000002   Commit ID
  1000000001   Commit ID
```

结合配置检查和修改的内容信息，使用"rollback configuration?"命令可以将配置恢复到先前的某个状态，XRv9k 支持最多回退到 100 个保存生效前的系统配置。

```
RP/0/0/CPU0: xrv1#rollback configuration ?
  last         Rollback last <n> commits made
  to           Rollback up to (and including) a specific commit
  to-exclude   Rollback up to (and excluding) a specific commit
  1000000003   Commit ID
  1000000002   Commit ID
  1000000001   Commit ID
```

XRv9k 路由器在启动时主要基于主用持久配置来恢复路由器的运行配置。主用持久配置与传统 Cisco IOS 中的启动配置不同，主用持久配置是基于提交基准点与绝对基准点创建的，它以系统数据库（SysDB）的格式进行保存，在路由器启动时将主用持久配置直接读取到系统内存中进而恢复启动前的配置和状态；而 IOS 的启动配置以 CLI 形式保存在文件中。

XRv9k 对每次成功的 commit 操作，都会创建一条 commit 数据库记录，基准点（refpoint）是这些数据库记录的索引，它是一组十位的数字，即上文中的 Commit ID。绝对基准点与提交基准点保存在配置文件系统中，默认路径为 XRv9k 路由器的"disk0:/config/"。每次对路由器进行配置修改和提交，都会导致绝对基准点与提交基准点的更新，其中绝对基准点是路由器定期更新的，提交基准点是每次 commit 操作时更新的，因此每次执行"commit"命令，就相当于将更改的配置生效并保存在配置文件系统中供路由器启动时恢复，而不再需要类似 IOS 中的"write"命令进行配置的单独保存。

2.4 CPSPT 五步实验法

本节分别从管理面、控制面和数据面开展相关实验。主要思路是通过管理面接口对协议进行配置；然后，通过报文捕获分析的方法得到交互的控制协议报文，了解协议格式封装及字段含义；进一步通过查看节点之间协议交互的状态，掌握协议交互流程及状态机操作；最后通过数据面发送特定的数据分组来验证端到端的可达性和控制面建立数据转发路径的正确性。上述操作概括起来包括 config、PktCapture、show、ping、traceroute，被称为 CPSPT 五步实验法。

- config

物理设备的配置工作主要通过 console 接口、SSH 或者 telnet 等通道进行设备管理和协议配置，本书中主要使用 EVE-NG 系统中提供的 Native console 或 HTML5 console 进行虚拟实验节点的配置。

- PktCapture

虚拟实验节点通信过程中的报文捕获工作主要由 Wireshark 软件完成，EVE-NG Lab 中节点虚拟接口的 Wireshark 调用由 EVE-NG 客户端集成软件包安装目录下的 win7_64bit_wireshark.reg 文件关联，而 Wireshark 自动化运行所需参数则由 wireshark_wrapper.bat 文件中的定义进行设置。

- show

Cisco IOS 中节点之间协议配置情况和交互的状态通过 show 命令进行查看，show 命令既可以查看本地记录的配置内容，也可以查看相关协议交互信息和生成表项。show 命令的使用和显示内容根据不同的命令行模式有不同的输出，我们可以使用"show？"查看当前模式能够显示的信息。

- ping

ping 命令主要用于验证虚拟实验节点的连通性，使用 ping 加上各类支持的参数，能够验证协议配置的正确性。注意 ping 命令的"mpls"参数代表发送以 MPLS 协议封装的 ping 包，这个功能在本书后面的章节中会多次用到。

- traceroute

在虚拟实验节点上面，使用 traceroute 命令侦测源节点到目的节点所经过的路由，验证节点的配置完成后数据是否按照实验预想的路径进行传输。traceroute 命令能够遍历数据包传输路径上的所有路由节点。traceroute 命令中同样支持"mpls"参数，使用"mpls"参数能够侦测到标签交换路径上每一个路由转发节点。

2.5 实验约定

为了便于读者结合配置信息更好地理解网络拓扑，本书对路由节点的路由器 ID 信息、接口编号以及接口 IP 地址等基础配置进行约定：每一个节点的名字采用 XRV-N 进行标识，该节点的路由器 ID 为 $N.N.N.N/32$，节点默认配置环回接口 0（Loopback0），Loopback0 接口的 IP 地址为 $N.N.N.N/32$。如果 XRV-N 与 XRV-M 存在互联接口，则 XRV-N 连接到 XRV-M 的接口名为 Gi0/0/0/M，接口 IP 地址配置为 $NM.1.1.N/24$，与之对应，XRV-M 连接到 XRV-N 的接口名为 Gi0/0/0/N，接口 IP 地址为 $NM.1.1.M/24$，其中 $N<M$。如果节点 XRV-N 存在一个没有邻居的接口，其 IP 地址配置为 $N.N.x.N/24$（x 为接口编号）。

在 PrefixSID 和 AdjSID 配置方面也进行约定。假定所有节点的 SR 全局块空间为 16000～23999，为 XRV-N 节点分配的 PrefixSID 为 $1600N$；SR 本地块空间为 24000～30000；LDP 的本地标签池为 $N00000$～$N99999$。

网络拓扑示例如图 2.36 所示。

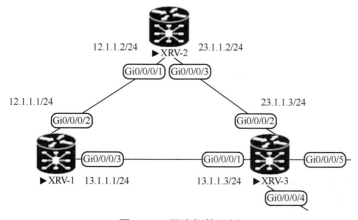

图 2.36　网络拓扑示例

以图 2.36 拓扑为例，路由器 XRV-2 的接口配置信息如下：

```
RP/0/0/CPU0:XRV-2(config)#show run
Building configuration...
!! IOS XR Configuration 6.3.1
!! Last configuration change at Wed Jan 27 13:39:00 2021 by cisco
!
hostname XRV-2
interface Loopback0
 ipv4 address 2.2.2.2 255.255.255.255
 no shutdown
!
interface GigabitEthernet0/0/0/1
 ipv4 address 12.1.1.2 255.255.255.0
 no shutdown
!
interface GigabitEthernet0/0/0/3
 ipv4 address 23.1.1.2 255.255.255.0
 no shutdown
!
router ospf 1
 router-id 2.2.2.2
 segment-routing mpls
 segment-routing forwarding mpls
 area 0
  interface Loopback0
   prefix-sid absolute 16002
  !
 !
!
segment-routing
 global-block 16000 23999
 local-block 24000 30000
!
mpls label range table 0 200000 299999
```

第 3 章

MPLS 技术原理与实战

3.1 MPLS 原理

MPLS 协议于 20 世纪 90 年代中期被提出，最初用于提高 IP 转发过程中最长匹配的查表效率，后来被应用于支持虚拟专用网（Virtual Private Network，VPN）以及服务质量（Quality of Service，QoS）等业务。MPLS 协议位于 OSI 七层模型中的第 2.5 层，即第二层（链路层）和第三层（IP 层）之间。

MPLS 报文头长度固定为 32bit，包括 20bit 的 MPLS 标签值（Label）、3bit 流分类（TC）、1bit 栈底标志（S）、8bit 生存时间（TTL），如图 3.1 所示。

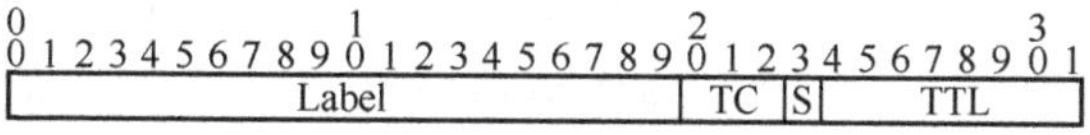

```
 0                   1                   2                   3
 0 1 2 3 4 5 6 7 8 9 0 1 2 3 4 5 6 7 8 9 0 1 2 3 4 5 6 7 8 9 0 1
                Label                    | TC |S|     TTL
```

图 3.1 MPLS 报文头格式

MPLS 架构由 RFC3031 定义，对于 MPLS 报文而言，转发操作主要有 3 种，包括压入（PUSH）、交换（SWAP）和弹出（POP）。

压入（PUSH）：当 IP 报文进入 MPLS 域时，MPLS 边界设备在报文二层首部和 IP 首部之间插入一个新标签；或者 MPLS 中间设备根据需要，在标签栈顶增加一个新的标签。

交换（SWAP）：当报文在 MPLS 域内转发时，根据 MPLS 转发表，用下一跳分配的标签替换 MPLS 报文的栈顶标签。

弹出（POP）：将 MPLS 报文栈顶标签去掉。

按照对 MPLS 的支持情况，可以将原来仅支持 IP 转发的路由器分为 IP 路由器和标签交换路由器（Label Switching Router，LSR）。LSR 通过相互连接构成了 MPLS 域，当 IP 报文进入 MPLS 域时，入口 LSR 执行 PUSH 操作，压入相应的 MPLS 标签，形成 MPLS 报文；MPLS 报文在 MPLS 域内经过 SWAP 操作持续转发；在 MPLS 域边界上，出口 LSR 执行 POP 操作，将 MPLS 报文还原为 IP 报文后按照 IP 报文进行转发。入口和出口 LSR 负责将 IP 报文封装为 MPLS 报文或者将 MPLS 报文还原为 IP 报文，这类 MPLS 域边界路由器又被称为标签边界路由器（Label Edge Router，LER）。

MPLS 域内部路由器的标签交换操作。LSR 接收到 MPLS 报文并进行必要的检查后，依据栈顶标签匹配本地 MPLS 转发表，按照 MPLS 转发表的操作类型进行标签交换。LSR 的 MPLS 转发表结构为{入标签，操作类型，出标签}。通过匹配入标签，获知操作类型，依据操作类型决定标签交换操作。出标签可以有一个或者多个，如果为一个，则相当于将 MPLS 报文的栈顶标签交换为出标签；如果为多个，则相当于将 MPLS 报文的栈顶标签弹出后又压入了标签，从而通过 SWAP 操作加深了标签栈的深度。

MPLS 域出口路由器的标签弹出操作。MPLS 域出口路由器作为标签交换路径上最后一跳，负责将 MPLS 报文还原为 IP 报文。LSR 接收 MPLS 报文，从 MPLS 报文提取 MPLS 标签值，依据标签值查找 MPLS 转发表，发现标签操作为 POP，于是弹出标签，还原为 IP 报文，送 IP 查表模块查表后转发。

对于最后一跳而言，有很多操作可以优化。一方面，从协议层面来讲，最后一跳路由器自身是能够获知自己为 MPLS 域出口路由器的，对于那些需要从 MPLS 域转发给 IP 域的 MPLS 报文而言，它无须查表，只需要弹出标签即可。因此，最后一跳路由器通常会为上游 LSR 分配特殊标签（0 或者 2），当收到携带特殊标签值的报文后，不需要查 MPLS 转发表，而直接执行标签弹出操作即可，这种特殊标签也被称为显式空标签。另一方面，最后一跳路由器在进行标签弹出还原为 IP 报文后可能还需要查 IP 表再进行转发，既然如此，可以在倒数第二跳就直接弹出标签还原为 IP 报文，这样在没有增加倒数第二跳路由器的操作复杂度的同时，可以使得最后一跳路由器无须执行标签弹出操作，进而提高转发效率。但倒数第二跳路由器并不能直接识别自己为倒数第二跳，因此，需要倒数第一跳路由器通过控制协议告知，告知的方法就是通告一个具有特殊含义的标签 3，这个标签被称为隐式空标签。如果 LSR 执行标签 SWAP 操作时，发现标签为 3，则直接弹出标签并转发给下一跳。

TTL 作为控制报文环路的一种方法，出现在 MPLS 报文头部。在将 IP 报文封装为 MPLS 报文，或者将 MPLS 报文还原为 IP 报文，或者在 MPLS 报文的嵌套压入和弹出过程中，都会涉及 TTL 值的操作。一种处理方法就是复制，即将 IP 报文的 TTL 值复制到 MPLS 的 TTL 域，在弹出 MPLS 标签时，将 MPLS 标签的 TTL 值重新复制到 IP 报文的 TTL 域，这种方式 IP 报文可以感知到 MPLS 域的存在。另一种处理方法就是重建，即在压入 MPLS 标签时为 TTL 设置初始值，在弹出 MPLS 标签时，MPLS 标签的 TTL 值不会复制到 IP 报文的 TTL 域，从 IP 报文的 TTL 域上来看，感知不到 MPLS 域的存在，虽然经过 MPLS 域内多个路由器的中继转发，但是 TTL 值没有发生变化。

MPLS 标签由路由器产生，并通过一定方式进行分发，包括静态、LDP、RSVP、BGP、IGP 等。下面通过实验来讲解采用静态配置标签和传统 LDP 的方式分发标签。一方面，通过静态标签配置实验，掌握基本的 MPLS 原理、MPLS 数据封装、静态配置方式；另一方面，通过动态标签分发协议，掌握标签分发原理、MPLS 标签的几种典型操作、标签表的查看指令以及整个过程中的 TTL 处理机制。通过对比实验，掌握静态和动态分配 MPLS 标签的区别和联系。

3.2 MPLS 静态配置

通过命令行为每个 FEC 分配标签，在每个节点建立起标签交换表项。MPLS 报文进入支持 MPLS 转发的路由器接口后，直接基于 MPLS 报文头部标签查找 MPLS 转发表，并实施标签操作后进行转发。静态配置的标签转发路径不是依赖于路由建立的。因此，静态配置的 MPLS 数据转发对路由协议没有依赖性。

3.2.1 实验环境

MPLS 标签交换拓扑如图 3.2 所示,所有节点已按第 2 章的实验约定配置了接口 IP 地址、环回接口地址等,未配置任何路由协议以及静态路由信息。下面通过静态配置方式为节点 5(即 XRV-5)的环回接口建立标签交换路径(Label Switched Path,LSP),要求从节点 1 去往该 FEC(5.5.5.5/32)的 LSP 为节点 1→2→3→4→5。

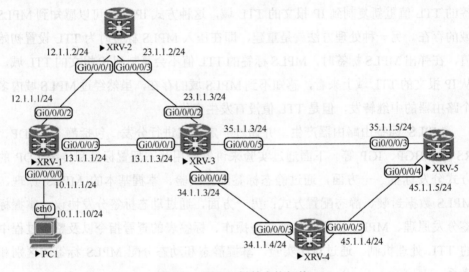

图 3.2 MPLS 标签交换拓扑

3.2.2 协议配置

按照实验要求,节点 1 为入口 LSR,实施标签压入操作;节点 5 为出口 LSR,实施标签弹出操作;节点 4 为倒数第二跳,该节点也可以实施标签弹出操作;其他节点进行标签交换操作。

静态标签表配置指令为:

```
mpls static address-family ipv4 unicast local-label <16-1048575>
allocate per-prefix <A.B.C.D/length> forward path <1-1> nexthop
<interface> <a.b.c.d> out-label [<16-1048575> | explicit-null | pop]
```

第 3 章 MPLS 技术原理与实战

约定在节点 N 上为 5.5.5.5/32 分配的标签为 NNNNNN，依次配置各节点的静态表项信息。例如，节点 2 的静态 MPLS 配置如下：

```
mpls static
 address-family ipv4 unicast
  local-label 222222 allocate per-prefix 5.5.5.5/32
   forward
    path 1 nexthop GigabitEthernet0/0/0/3 23.1.1.3 out-label 333333
```

节点 2 为该 FEC 分配标签 222222，标签操作为交换，下一跳为节点 3，出标签为 333333，该出标签正是下一跳节点 3 为该 FEC 分配的入标签。配置完后可以查看本地 MPLS 转发表项信息：

```
RP/0/0/CPU0:XRV-2#show mpls forwarding detail
Sat Jan 30 09:31:46.048 UTC
Local    Outgoing    Prefix           Outgoing      Next Hop       Bytes
Label    Label       or ID            Interface                    Switched
------   ----------  ---------------  ------------  ------------   ------------
222222   333333      5.5.5.5/32       Gi0/0/0/3     23.1.1.3       81688
     Updated: Jan 30 01:01:24.805
     Version: 5, Priority: 4
     Label Stack (Top -> Bottom): { 333333 }
     NHID: 0x0, Encap-ID: N/A, Path idx: 0, Backup path idx: 0,
Weight: 0
     MAC/Encaps: 14/18, MTU: 1500
     Outgoing Interface: GigabitEthernet0/0/0/3 (ifhandle 0x00000080)
     Packets Switched: 815
```

如果 LSP 采用显式空标签，则节点 4 的配置为：

```
mpls static
 address-family ipv4 unicast
  local-label 444444 allocate per-prefix 5.5.5.5/32
   forward
    path 1 nexthop GigabitEthernet0/0/0/5 45.1.1.5 out-label
explicit-null
```

对应的 MPLS 转发表项信息如下，可以看到已经建立了入标签为 444444、

45

出标签为 0 的 MPLS 转发表项。

```
RP/0/0/CPU0:XRV-4#show mpls forwarding detail
Sat Jan 30 09:35:05.394 UTC
Local  Outgoing    Prefix        Outgoing       Next Hop      Bytes
Label  Label       or ID         Interface                    Switched
------ ----------- ------------- -------------- ------------- ------------
444444 Exp-Null-v4 5.5.5.5/32    Gi0/0/0/5      45.1.1.5      0
    Updated: Jan 30 09:39:35.605
    Version: 11, Priority: 4
    Label Stack (Top -> Bottom): { 0 }
    NHID: 0x0, Encap-ID: N/A, Path idx: 0, Backup path idx: 0,
Weight: 0
    MAC/Encaps: 14/18, MTU: 1500
    Outgoing Interface: GigabitEthernet0/0/0/5 (ifhandle 0x000000e0)
    Packets Switched: 0
```

如果 LSP 采用倒数第二跳弹出，则节点 4 配置为：

```
mpls static
 address-family ipv4 unicast
  local-label 444444 allocate per-prefix 5.5.5.5/32
   forward
    path 1 nexthop GigabitEthernet0/0/0/5 45.1.1.5 out-label pop
```

对应的 MPLS 转发表项信息如下，此时入标签 444444 被弹出，标签栈信息显示为 Imp-Null，即隐式空标签。

```
RP/0/0/CPU0:XRV-4#show mpls forwarding detail
Sat Jan 30 09:35:05.394 UTC
Local  Outgoing    Prefix        Outgoing       Next Hop      Bytes
Label  Label       or ID         Interface                    Switched
------ ----------- ------------- -------------- ------------- ------------
444444 Pop         5.5.5.5/32    Gi0/0/0/5      45.1.1.5      3360
    Updated: Jan 30 03:17:30.116
    Version: 10, Priority: 4
    Label Stack (Top -> Bottom): { Imp-Null }
```

```
        NHID: 0x0, Encap-ID: N/A, Path idx: 0, Backup path idx: 0,
Weight: 0
        MAC/Encaps: 14/14, MTU: 1500
        Outgoing Interface: GigabitEthernet0/0/0/5 (ifhandle 0x000000e0)
        Packets Switched: 35
```

在支持 MPLS 转发功能的节点上开启 MPLS 转发后，将会自动启动特殊标签（标签值为 0～16）处理的功能。对于标签 0 而言，其操作就是弹出标签并进行 IP 路由。因此，节点 5 作为 LSP 的出口，不需要进行 MPLS 转发表项配置。

如果 LSP 采用显式空标签，则 MPLS 报文到达节点 4 后，栈顶标签被交换为 0，此时，节点 5 需要开启 MPLS 转发功能，这样它在收到入标签为 0 的 MPLS 报文后，将直接执行标签弹出操作，而不需要查找 MPLS 转发表。

如果 LSP 采用倒数第二跳弹出，则 MPLS 报文到达节点 4 后，标签被弹出，MPLS 报文还原为 IP 报文，此时，节点 5 将收到 IP 报文，按照 IP 报文进行处理即可，节点 5 不需要开启 MPLS 转发功能。

3.2.3　使能 MPLS 转发功能

默认情况下 MPLS 转发功能并不开启，也就是虽然已经配置 MPLS 转发表，但路由节点还不具备处理 MPLS 报文转发的能力。

MPLS 转发功能的开启与标签生成方式相关。MPLS 转发功能是与接口关联的，管理员按需在路由器的部分接口上开启 MPLS 转发功能。只需要在 MPLS 报文入接口启动 MPLS 转发即可，MPLS 报文的出接口不需要配置。例如图 3.2，如果仅建立单向 LSP（节点 1→2→3→4→5），那么只需要在节点 2 的接口 Gi0/0/0/1、节点 3 的接口 Gi0/0/0/2、节点 4 的接口 Gi0/0/0/3、节点 5 的接口 Gi0/0/0/4 上开启 MPLS 转发功能即可。

针对静态标签配置方式，使能接口 MPLS 转发能力的命令为：
```
mpls static interface <intf-name>
```

例如，节点 3 接口的静态 MPLS 配置如下：
```
mpls static
 interface GigabitEthernet0/0/0/2
 address-family ipv4 unicast
```

```
local-label 333333 allocate per-prefix 5.5.5.5/32
 forward
  path 1 nexthop GigabitEthernet0/0/0/4 34.1.1.4 out-label 444444
```

3.3 MPLS 数据转发

在完成节点 1～节点 5 的 MPLS 转发表项配置，开启接口的 MPLS 转发功能后，对静态 MPLS 的配置已经完成。通过 MPLS 的操作管理和维护（Operation Administration and Maintenance，OAM）能力可以检测 LSP，快速定位路径建立问题。下面使用 MPLS OAM 功能验证 MPLS 路径的正确性。

与 IP 追踪定位故障方法类似，MPLS OAM 提供了 MPLS ping 和 MPLS traceroute 两种模式。前者用于检查入口和出口之间 LSR 的连通性，定位与目的节点之间的 LSP 是否正常建立；如果没有正常建立，通过反馈的原因码等信息，初步确定 LSP 失效的原因。后者可以清晰定位 LSP 经过的每一个节点，以及该节点为 LSP 分配的标签信息、下游标签映射信息等，对 MPLS 路径进行更为精细的追踪分析与故障定位。

3.3.1 MPLS Echo 模式

MPLS ping/traceroute 都使用 MPLS Echo Request 和 MPLS Echo Reply 机制，两者仅在 Request 和 Reply 的信息内容上存在差异，RFC8029 对 Echo 模式报文进行了详细定义。

MPLS Echo 报文按照要求在 IP 报文前封装一层或者多层 MPLS 标签。MPLS Echo 报文承载的 IP 报文具有一定的特殊性，其目的 IP 地址为 127.x.x.x 而不是 FEC 地址。MPLS Echo 报文在转发过程中，可能因为未知原因将 MPLS 标签弹出，如果目的 IP 地址为 FEC 地址，则该 MPLS 探测包在变为 IP 报文后会沿着 IP 转发路径继续转发，这样就实现不了 MPLS 路径探测的目的。在将 IP 地址设置为 127.x.x.x 后，将标签弹出还原为 IP 报文的节点，在依据目的 IP 地址进行路由时，发现目的 IP 为本地环回地址，于是会将该 Echo 报文交给控制面对应的应用程序进行处理。

MPLS Echo 承载的 IP 报文 TTL 值必须设置为 1，且携带路由器告警（Router Alert）选项，通过 UDP 3503 端口进行协议交互，MPLS Echo Request 报文示例如图 3.3 所示。

```
11 3497.452441   12.1.1.1   127.0.0.1   MPLS ECHO   114 MPLS Echo Request
> Frame 11: 114 bytes on wire (912 bits), 114 bytes captured (912 bits) on interface -, id 0
> Ethernet II, Src: 50:00:00:01:00:03 (50:00:00:01:00:03), Dst: 50:00:00:02:00:02 (50:00:00:02:00:02)
v MultiProtocol Label Switching Header, Label: 222222, Exp: 0, S: 1, TTL: 255
    0011 0110 0100 0000 1110 .... .... .... = MPLS Label: 222222
    .... .... .... .... .... 000. .... .... = MPLS Experimental Bits: 0
    .... .... .... .... .... ...1 .... .... = MPLS Bottom Of Label Stack: 1
    .... .... .... .... .... .... 1111 1111 = MPLS TTL: 255
v Internet Protocol Version 4, Src: 12.1.1.1, Dst: 127.0.0.1
    0100 .... = Version: 4
    .... 0110 = Header Length: 24 bytes (6)
  > Differentiated Services Field: 0x00 (DSCP: CS0, ECN: Not-ECT)
    Total Length: 96
    Identification: 0x0009 (9)
  > Flags: 0x40, Don't fragment
    Fragment Offset: 0
  > Time to Live: 1
    Protocol: UDP (17)
    Header Checksum: 0x587d [validation disabled]
    [Header checksum status: Unverified]
    Source Address: 12.1.1.1
    Destination Address: 127.0.0.1
  > Options: (4 bytes), Router Alert
> User Datagram Protocol, Src Port: 3503, Dst Port: 3503
v Multiprotocol Label Switching Echo
    Version: 1
  > Global Flags: 0x0000
    Message Type: MPLS Echo Request (1)
    Reply Mode: Reply via an IPv4/IPv6 UDP packet (2)
    Return Code: No return code (0)
    Return Subcode: 0
    Sender's Handle: 0x00001ebb
    Sequence Number: 5
    Timestamp Sent: Jan 31, 2021 02:00:33.794193984 UTC
    Timestamp Received: (0)Jan  1, 1970 00:00:00.000000000 UTC
  > Vendor Private
  > Target FEC Stack
```

图 3.3　MPLS Echo Request 报文示例

IP Echo 报文通过因特网控制消息协议（Internet Control Message Protocol，ICMP）承载，由协议栈完成 Echo 报文的处理；MPLS Echo 报文则需要独立的应用程序来解析、处理和响应。MPLS Echo 应用程序会在本地协议栈 3503 端口监听，当 MPLS 报文被还原为 IP 报文后，因为目的 IP 为环回接口而将 Echo 报文送到控制面协议栈，由控制面协议栈的应用程序对 Echo 报文进行处理。默认情况下，节点并不开启对 MPLS Echo 报文的处理，此时，如果 MPLS Echo 报文到达控制面后，因为没有应用程序在 UDP 3503 端口上监听，所以协议栈可能会回复“目的端口不可达”的 ICMP 响应。要开启 MPLS Echo 报文的处理功能，需要在全局配置模式下输入：

```
mpls oam
```

3.3.2 MPLS Echo 报文封装机制

MPLS Echo 报文有两种封装机制，一种为基于 Prefix-FEC 生成标签栈，另一种为基于 Nil-FEC 直接生成标签栈。

（1）基于 Prefix-FEC 生成标签栈

基于 Prefix-FEC 生成标签栈的指令如下：

```
ping mpls ipv4 A.B.C.D/length fec-type [ generic | ldp | bgp ]
```

当前版本需要指定 fec-type。fec-type 用于填充 MPLS Echo 报文中的 Target FEC Stack 字段，RFC8029 和 RFC8087 定义了该 Sub-TLV 的相关信息，见表3.1。

表 3.1 Target FEC Sub-TLV 的相关信息

子类型	值域描述
1	LDP IPv4 prefix
2	LDP IPv6 prefix
3	RSVP IPv4 LSP
4	RSVP IPv6 LSP
6	VPN IPv4 prefix
7	VPN IPv6 prefix
8	L2 VPN endpoint
9	"FEC 128" Pseudowire-IPv4(deprecated)
10	"FEC 128" Pseudowire-IPv4
11	"FEC 129" Pseudowire-IPv4
12	BGP labeled IPv4 prefix
13	BGP labeled IPv6 prefix
14	Generic IPv4 prefix
15	Generic IPv6 prefix
16	Nil FEC
24	"FEC 128" Pseudowire - IPv6
25	"FEC 129" Pseudowire - IPv6
34	IPv4 IGP-Prefix Segment ID
35	IPv6 IGP-Prefix Segment ID
36	IGP-Adjacency Segment ID

举例而言，当输入"ping mpls ipv4 5.5.5.5/32 fec-type generic"指令后，可以看到，指令关键字 ipv4 和 fec-type 中的字段被填充到 MPLS Echo 报文 Target FEC Stack 的对应域中，如图 3.4 所示。

```
> MultiProtocol Label Switching Header, Label: 222222, Exp: 0, S: 1, TTL: 255
> Internet Protocol Version 4, Src: 12.1.1.1, Dst: 127.0.0.1
> User Datagram Protocol, Src Port: 3503, Dst Port: 3503
∨ Multiprotocol Label Switching Echo
    Version: 1
  > Global Flags: 0x0000
    Message Type: MPLS Echo Request (1)
    Reply Mode: Reply via an IPv4/IPv6 UDP packet (2)
    Return Code: No return code (0)
    Return Subcode: 0
    Sender's Handle: 0x00001ebf
    Sequence Number: 2
    Timestamp Sent: Jan 31, 2021 02:23:04.511663499 UTC
    Timestamp Received: (0)Jan  1, 1970 00:00:00.000000000 UTC
  > Vendor Private
  ∨ Target FEC Stack
      Type: Target FEC Stack (1)
      Length: 12
    ∨ FEC Element 1: Generic IPv4 prefix
        Type: Generic IPv4 prefix (14)
        Length: 5
        IPv4 Prefix: 5.5.5.5
        Prefix Length: 32
        Padding: 000000
```

图 3.4 Target FEC Stack 示例

需要注意的是，基于 FEC 的标签封装模式对本地路由存在依赖，也就是说如果本地路由表中不存在到达该 FEC 的路由，那么 MPLS Echo 报文是无法发送的。由于当前实验环境下并未运行任何路由协议，所以节点 1 不会自动学习到达 5.5.5.5/32 的路由，为了能够发送 MPLS Echo 报文，需要针对 5.5.5.5/32 配置一条静态路由。此时节点 1 的配置如下：

```
router static
 address-family ipv4 unicast
  5.5.5.5/32 GigabitEthernet0/0/0/3 13.1.1.3
 !
!
mpls oam
!
mpls static
```

```
interface GigabitEthernet0/0/0/2
interface GigabitEthernet0/0/0/3
address-family ipv4 unicast
 local-label 111111 allocate per-prefix 5.5.5.5/32
  forward
   path 1 nexthop GigabitEthernet0/0/0/2 12.1.1.2 out-label 222222
```

这里故意将静态路由 5.5.5.5/32 的下一跳与出接口配置得与 MPLS 转发表的下一跳及出接口不一致。虽然两者不一致，只要本地存在有效的路由，MPLS Echo 报文就能发送出去，而且是按照 MPLS 转发表从 Gi0/0/0/2 发送给节点 2。

（2）基于 Nil-FEC 直接生成标签栈

```
ping mpls nil-fec labels {label [,label]} [output {interfacetx-interface} [nexthop nexthop-ip-addr] ]
```

对于均具有特殊含义的保留标签而言，这些标签不与具体 FEC 建立映射关系，但有时需要压入这种标签以辅助进行故障诊断。Nil-FEC 提供了一种直接填写任意标签栈的方法。

通过 Nil-FEC 可以直接指定发出的 MPLS Echo 报文的 MPLS 头部标签。基于 Nil-FEC 产生的 MPLS Echo 不需要查找本地路由和 MPLS 转发表，而是依据配置的下一跳解析并封装 MPLS Echo 目的 MAC 地址，依据配置的出接口填写源 IP 地址和源 MAC 地址，并最终将该 MPLS Echo 报文从指定的出接口发送。

举例而言，在节点 1 输入 "ping mpls nil-fec labels 0,1,2,3,4,5,6 output interface gi0/0/0/3 nexthop 13.1.1.3" 指令后，可以在其 Gi0/0/0/3 接口上捕获 MPLS Echo 报文，如图 3.5 所示，共压入 7 层标签，标签 3 为隐式空标签，仅在出口节点通过标签分发协议通告倒数第二跳弹出时使用，不会出现在 MPLS 数据面，所以，不会向数据包中压入标签 3。除了配置的标签，系统默认增加一个显式空标签作为栈底标签，确保在倒数第二跳弹出条件下，仍然以 MPLS 封装格式送到 LSP 出口节点。MPLS Echo 报文的源 IP 地址从节点 1 的出接口 Gi0/0/0/3 上获取，Echo 报文中 Target FEC Stack 显示子类型为 Nil FEC，标签为系统默认添加的显式空标签。

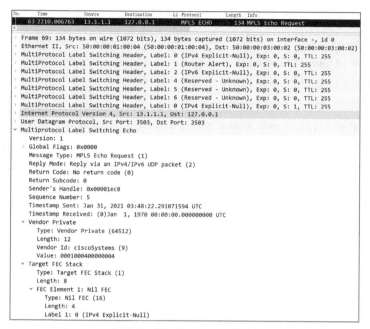

图 3.5 Nil FEC Stack 示例

3.3.3 MPLS ping

下面用 MPLS ping 指令在节点 1 上通过 Prefix-FEC 方式验证去往节点 5 的 LSP 的可达性。

采用图 3.2 拓扑进行实验，所有节点针对 5.5.5.5/32 配置 MPLS 转发表项，LSP 为节点 1→2→3→4→5，最后一跳进行标签弹出。各节点开启了 MPLS OAM 功能，但未配置任何动态路由协议，仅在节点 1 上针对 Prefix-FEC 配置了一条静态路由。各节点协议配置信息如下：

```
XRV-1:
router static
 address-family ipv4 unicast
  5.5.5.5/32 GigabitEthernet0/0/0/3 13.1.1.3
 !
!
mpls oam
```

```
!
mpls static
 interface GigabitEthernet0/0/0/2
 interface GigabitEthernet0/0/0/3
 address-family ipv4 unicast
  local-label 111111 allocate per-prefix 5.5.5.5/32
   forward
    path 1 nexthop GigabitEthernet0/0/0/2 12.1.1.2 out-label 222222
```

XRV-2：
```
mpls oam
!
mpls static
 interface GigabitEthernet0/0/0/1
 interface GigabitEthernet0/0/0/3
 address-family ipv4 unicast
  local-label 222222 allocate per-prefix 5.5.5.5/32
   forward
    path 1 nexthop GigabitEthernet0/0/0/3 23.1.1.3 out-label 333333
```

XRV-3：
```
mpls oam
!
mpls static
 interface GigabitEthernet0/0/0/1
 interface GigabitEthernet0/0/0/2
 interface GigabitEthernet0/0/0/4
 interface GigabitEthernet0/0/0/5
 address-family ipv4 unicast
  local-label 333333 allocate per-prefix 5.5.5.5/32
   forward
    path 1 nexthop GigabitEthernet0/0/0/4 34.1.1.4 out-label 444444
```

XRV-4：
```
mpls oam
!
mpls static
```

第 3 章　MPLS 技术原理与实战

```
 interface GigabitEthernet0/0/0/3
 interface GigabitEthernet0/0/0/5
 address-family ipv4 unicast
  local-label 444444 allocate per-prefix 5.5.5.5/32
   forward
    path 1 nexthop GigabitEthernet0/0/0/5 45.1.1.5 out-label
explicit-null
```

XRV-5：

```
mpls oam
!
mpls static
 interface GigabitEthernet0/0/0/3
 interface GigabitEthernet0/0/0/4
!
```

通过 MPLS ping 指令"ping mpls ipv4 5.5.5.5/32 fec-type generic"，验证到 5.5.5.5/32 的 LSP 是否可达。在节点 5 的 Gi0/0/0/4 接口上进行抓包，节点 5 收到的 MPLS Echo Request 报文如图 3.6 所示，可以看到，节点 4 已经按照 MPLS 转发表将 MPLS ping 的栈顶标签交换为显式空标签发给节点 5，TTL 值也从初始 255 逐跳减少至 252（在节点 2、节点 3、节点 4 上依次减 1），但是节点 5 并未响应该 MPLS Echo 请求。

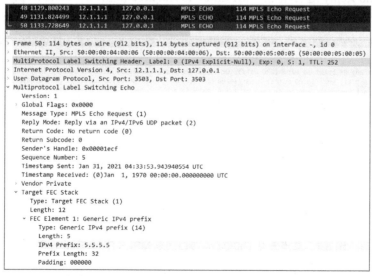

图 3.6　节点 5 收到的 MPLS Echo Request 报文

从节点 5 的配置来看，节点并未配置任何 MPLS 转发表项信息，仅在接口上开启了 MPLS 转发功能。那么节点 5 作为最后一跳是否必须配置 MPLS 转发表项呢？答案是不需要。

节点 5 的接口已经开启了 MPLS 转发功能，它必然按照 MPLS 的处理逻辑接收该 MPLS Echo Request 报文，根据对显式空标签的操作流程，弹出该标签还原为 IP 报文。由于 IP 报文的目的 IP 地址为 127.0.0.1，因此将该报文按照目的端口号交付给控制面对应的应用进程 MPLS OAM 进行处理。如果本地启动了 MPLS OAM 服务，它将处理 MPLS Echo Request 报文，从中提取 Target FEC，验证本地是否为该 FEC 的 LSP 尾端节点，如果是则相应产生 MPLS Echo Reply 报文，报文的目的 IP 地址自然就是 12.1.1.1。节点 5 依据目的 IP 地址 12.1.1.1 查找本地路由，决定从哪个接口将 Reply 报文发送出去。本例中由于节点 5 没有到达 12.1.1.1 的路由，所以 MPLS Echo Reply 报文无法发送出去。

为了验证这一点，我们在节点 5 上配置 12.1.1.0/24 路由，下一跳指向节点 4：

```
router static
 address-family ipv4 unicast
  12.1.1.0/24 GigabitEthernet0/0/0/4 45.1.1.4
```

可以看到节点 5 能够正确响应此 MPLS Echo Request，并通过 UDP 封装的 MPLS Echo Reply 进行响应，MPLS Echo Request 和 MPLS Echo Reply 在同一个接口上收发，节点 5 从 Gi0/0/0/4 接口回复 MPLS Echo Reply 报文，如图 3.7 所示。

图 3.7　节点 5 从 Gi0/0/0/4 接口回复 MPLS Echo Reply 报文

或者，在节点 5 上配置 12.1.1.0/24 路由，下一跳指向节点 3：

```
router static
 address-family ipv4 unicast
  12.1.1.0/24 GigabitEthernet0/0/0/3 35.1.1.3
```

此时需要在节点 5 的 Gi0/0/0/3 接口进行抓包。节点 5 的 Gi0/0/0/4 接口收到 MPLS Echo Request 报文后，MPLS Echo Reply 通过 Gi0/0/0/3 接口发送。节点 5 从 Gi0/0/0/3 接口发送 MPLS Echo Reply 报文如图 3.8 所示，源 IP 地址为接收 MPLS Echo Request 报文的接口 IP 地址，并没有因为从 Gi0/0/0/3 接口响应而选择使用该接口地址作为源 IP 地址。

```
    4 5.706178      45.1.1.5      12.1.1.1         MPLS ECHO        90 MPLS Echo Reply
    5 7.596509      45.1.1.5      12.1.1.1         MPLS ECHO        90 MPLS Echo Reply
Frame 5: 90 bytes on wire (720 bits), 90 bytes captured (720 bits) on interface -, id 0
Ethernet II, Src: 50:00:00:05:00:04 (50:00:00:05:00:04), Dst: 50:00:00:03:00:06 (50:00:00:03:00:06)
Internet Protocol Version 4, Src: 45.1.1.5, Dst: 12.1.1.1
User Datagram Protocol, Src Port: 3503, Dst Port: 3503
Multiprotocol Label Switching Echo
    Version: 1
  › Global Flags: 0x0000
    Message Type: MPLS Echo Reply (2)
    Reply Mode: Reply via an IPv4/IPv6 UDP packet (2)
    Return Code: Replying router is an egress for the FEC at stack depth RSC (3)
    Return Subcode: 1
    Sender's Handle: 0x00001ed6
    Sequence Number: 5
    Timestamp Sent: Jan 31, 2021 06:36:49.478681814 UTC
    Timestamp Received: Jan 31, 2021 06:36:45.078914714 UTC
  › Vendor Private
```

图 3.8 节点 5 从 Gi0/0/0/3 接口发送 MPLS Echo Reply 报文

在节点 2、3、4、5 上依次配置 12.1.1.0/24 的静态路由，确保该路由的静态路径为节点 5→4→3→2→1。

完成关于 12.1.1.0/24 的静态路由配置后，在节点 1 上执行 MPLS ping 指令"ping mpls ipv4 5.5.5.5/32 fec-type generic verbose"，得到如下响应结果，结果显示 MPLS ping 操作正常。

```
RP/0/0/CPU0:XRV-1#ping mpls ipv4 5.5.5.5/32 fec-type generic verbose

Sending 5, 100-byte MPLS Echos to 5.5.5.5/32,
      timeout is 2 seconds, send interval is 0 msec:

Codes: '!' - success, 'Q' - request not sent, '.' - timeout,
  'L' - labeled output interface, 'B' - unlabeled output interface,
  'D' - DS Map mismatch, 'F' - no FEC mapping, 'f' - FEC mismatch,
```

```
'M' - malformed request, 'm' - unsupported tlvs, 'N' - no rx label,
'P' - no rx intf label prot, 'p' - premature termination of LSP,
'R' - transit router, 'I' - unknown upstream index,
'X' - unknown return code, 'x' - return code 0

Type escape sequence to abort.

!     size 100, reply addr 45.1.1.5, return code 3
!     size 100, reply addr 45.1.1.5, return code 3
!     size 100, reply addr 45.1.1.5, return code 3
!     size 100, reply addr 45.1.1.5, return code 3
!     size 100, reply addr 45.1.1.5, return code 3
```

MPLS ping 提供了丰富的控制参数，以满足不同场景需求，相关配置参数如下：

```
ping mpls ipv4 A.B.C.D/length [exp <0-7>] [ttl <1-255>] [source
A.B.C.D] [fec-type [generic|ldp|bgp]] [interval <0-3600000>] [repeat
<1-2147483647>] [size <100-17986>] [pad <0-ffff>] [reply (dscp <0-63> |
mode [ipv4|no-reply|router-alert]) ]
```

3.3.4　MPLS traceroute

MPLS traceroute 为源节点提供了查询 LSP 上每一个节点转发表项信息的能力。MPLS traceroute 指令如下：

```
traceroute mpls ipv4 A.B.C.D/length fec-type [ generic | ldp | bgp ]
```

LSP 的中间节点在收到 MPLS traceroute 报文后，将本地 MPLS 转发表信息封装到响应报文的 Downstream Mapping TLV 中，回复给源节点。相关信息包括连接下游节点的接口最大传输单元（Maximum Transmission Unit，MTU）、多径信息、下游节点地址与接口地址以及对应标签栈信息。

MPLS traceroute 功能是使用 MPLS Echo Request 和 MPLS Echo Reply 报文封装相关信息。与 MPLS ping 的不同，MPLS traceroute 在 Echo 报文中需要携带 Downstream Mapping TLV（类型 2）。按照 RFC8029 规定，当前已经废弃了

Downstream Mapping TLV，而采用新定义的 Downstream Detailed Mapping TLV（类型 20）来承载 traceroute 相关信息。但在 Cisco IOS XR（Version 6.3.1）中还使用 Downstream Mapping TLV，而不支持 Downstream Detailed Mapping TLV。

在确保 5.5.5.5/32 的静态 LSP 为节点 1→2→3→4→5、12.1.1.0/24 的静态路径为节点 5→4→3→2→1 的条件下，在节点 1 上执行指令"traceroute mpls ipv4 5.5.5.5/32 fec-type generic verbose"，得到如下 traceroute 结果。

```
RP/0/0/CPU0:XRV-1#traceroute mpls ipv4 5.5.5.5/32 fec-type generic
verbose

Tracing MPLS Label Switched Path to 5.5.5.5/32, timeout is 2 seconds

Codes: '!' - success, 'Q' - request not sent, '.' - timeout,
  'L' - labeled output interface, 'B' - unlabeled output interface,
  'D' - DS Map mismatch, 'F' - no FEC mapping, 'f' - FEC mismatch,
  'M' - malformed request, 'm' - unsupported tlvs, 'N' - no rx label,
  'P' - no rx intf label prot, 'p' - premature termination of LSP,
  'R' - transit router, 'I' - unknown upstream index,
  'X' - unknown return code, 'x' - return code 0

Type escape sequence to abort.

  0 12.1.1.1 12.1.1.2 MRU 1500 [Labels: 222222 Exp: 0]
L 1 12.1.1.2 23.1.1.3 MRU 1500 [Labels: 333333 Exp: 0] 10 ms, ret code 8
L 2 23.1.1.3 34.1.1.4 MRU 1500 [Labels: 444444 Exp: 0] 10 ms, ret code 8
L 3 34.1.1.4 45.1.1.5 MRU 1500 [Labels: explicit-null Exp: 0]
10 ms, ret code 8
! 4 45.1.1.5 10 ms, ret code 3
```

从结果可以获知，去往 5.5.5.5/32 第一跳 IP 地址为 12.1.1.2，其出标签为 333333，下游 IP 地址为 23.1.1.3；第二跳为 23.1.1.3，其出标签为 444444，下游 IP 地址为 34.1.1.4；第三跳为 34.1.1.4，其出标签为 explicit-null，下游 IP 地址为 45.1.1.5；第四跳为 45.1.1.5，它就是 5.5.5.5/32 LSP 的终点。

在上述过程中，在节点 1 的 Gi0/0/0/2 接口进行抓包分析，节点 1 执行 MPLS

traceroute 过程中交互的报文如图 3.9 所示，整个过程中，节点 1 共发送 4 个 Echo Request，得到了 4 个 Echo Reply，响应报文的源 IP 地址均为各节点接收 Echo Request 请求的入接口 IP 地址。

No.	Time	Source	Destination	Li	Protocol	Length Info
1	0.000000	12.1.1.1	127.0.0.1		MPLS ECHO	138 MPLS Echo Request
2	0.001445	12.1.1.2	12.1.1.1		MPLS ECHO	114 MPLS Echo Reply
3	0.005969	12.1.1.1	127.0.0.1		MPLS ECHO	138 MPLS Echo Request
4	0.009966	23.1.1.3	12.1.1.1		MPLS ECHO	114 MPLS Echo Reply
5	0.014043	12.1.1.1	127.0.0.1		MPLS ECHO	138 MPLS Echo Request
6	0.020497	34.1.1.4	12.1.1.1		MPLS ECHO	114 MPLS Echo Reply
7	0.025117	12.1.1.1	127.0.0.1		MPLS ECHO	138 MPLS Echo Request
8	0.034167	45.1.1.5	12.1.1.1		MPLS ECHO	90 MPLS Echo Reply

图 3.9　节点 1 执行 MPLS traceroute 过程中交互的报文

以第二个 MPLS Echo Request 报文为例，节点 1 发送的 TTL=2 的 MPLS Echo Request 报文如图 3.10 所示，节点 1 设置该 MPLS 报文 TTL=2，携带 Target FEC Stack，告知 TTL 超时节点应该查询本地哪个 FEC 的标签转发信息，并进一步携带上一次查询反馈（节点 2 反馈）的标签转发信息，以便核对 TTL 超时的节点入口信息的正确性。

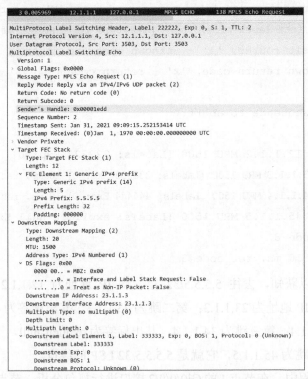

图 3.10　节点 1 发送的 TTL=2 的 MPLS Echo Request 报文

第二个 MPLS Echo Request 报文到达节点 3 后因为 TTL 值被减为 0，触发节点 3 的 TTL 超时处理，它从 Target FEC Stack 中提取对应的 Prefix 信息，核对 Downstream Mapping 信息后，检索本地 MPLS 转发表，将本地 MPLS 转发表按要求封装后反馈，节点 3 发送的 MPLS Echo Reply 报文如图 3.11 所示。通过该响应报文，节点 1 获知节点 3 去往 5.5.5.5/32 的下一跳为 34.1.1.4，出标签为 444444，这个信息将被节点 1 保留并复制到下一个 MPLS Echo Request 报文中，协助节点 4 核对入口信息的正确性。

```
4 0.009966      23.1.1.3      12.1.1.1       MPLS ECHO      114 MPLS Echo Reply
Frame 4: 114 bytes on wire (912 bits), 114 bytes captured (912 bits) on interface -, id 0
Ethernet II, Src: 50:00:00:02:00:02 (50:00:00:02:00:02), Dst: 50:00:00:01:00:03 (50:00:00:01:00:03)
Internet Protocol Version 4, Src: 23.1.1.3, Dst: 12.1.1.1
User Datagram Protocol, Src Port: 3503, Dst Port: 3503
Multiprotocol Label Switching Echo
    Version: 1
  > Global Flags: 0x0000
    Message Type: MPLS Echo Reply (2)
    Reply Mode: Reply via an IPv4/IPv6 UDP packet (2)
    Return Code: Label switched at stack-depth RSC (8)
    Return Subcode: 1
    Sender's Handle: 0x00001edd
    Sequence Number: 2
    Timestamp Sent: Jan 31, 2021 09:09:15.252153414 UTC
    Timestamp Received: Jan 31, 2021 09:09:08.692534274 UTC
  > Vendor Private
  ∨ Downstream Mapping
      Type: Downstream Mapping (2)
      Length: 20
      MTU: 1500
      Address Type: IPv4 Numbered (1)
    ∨ DS Flags: 0x00
        0000 00.. = MBZ: 0x00
        .... ..0. = Interface and Label Stack Request: False
        .... ...0 = Treat as Non-IP Packet: False
      Downstream IP Address: 34.1.1.4
      Downstream Interface Address: 34.1.1.4
      Multipath Type: no multipath (0)
      Depth Limit: 0
      Multipath Length: 0
    ∨ Downstream Label Element 1, Label: 444444, Exp: 0, BOS: 1, Protocol: 0 (Unknown)
        Downstream Label: 444444
        Downstream Exp: 0
        Downstream BOS: 1
        Downstream Protocol: Unknown (0)
```

图 3.11　节点 3 发送的 MPLS Echo Reply 报文

3.4　MPLS TTL 传播机制

MPLS 报文的 TTL 与其承载的 IP 报文的 TTL 可能存在关联性，关联发生在

IP 报文进入 MPLS 域和离开 MPLS 域的时刻。前者是指在 IP 报文头部加入 MPLS 标签时，MPLS 标签的 TTL 可以设置为默认值，或者将 IP 头部的 TTL 值复制到 MPLS 的 TTL 域；后者是指将 MPLS 标签弹出时，IP 头部的 TTL 值可以保持不变，也可以将弹出 MPLS 标签的 TTL 值复制到 IP 头部的 TTL 域。

按照 MPLS 标签的 TTL 值是否从 IP 头中复制将 TTL 传播分为两种模式，一种模式称为统一模式，即压入或者弹出标签时都需要复制操作；另一种模式称为管道模式，即压入或弹出时不在 IP 和 MPLS 头部进行 TTL 复制操作。从端到端数据传送来看，统一模式下，报文经过 IP 路由器或者 LSR 转发时，对 TTL 进行统一的操作，终端能够从 TTL 的变化感知 MPLS 域的存在；在管道模式下，整个 MPLS 域被视为一个管道，TTL 不会随 LSR 的数据转发而逐跳改变。

实验环境中 IOS 默认采用统一模式：即压入 MPLS 标签时，其 TTL 值从 IP 头部复制；在弹出 MPLS 标签时，将 MPLS 标签的 TTL 值复制到 IP 头部。统一模式真实反映了数据报文经过的路由器跳数。实际运行过程中，可以依据需求选择在压入标签时关闭复制功能，或者在弹出标签时关闭复制功能。设置 TTL 复制的指令如下：

```
mpls ip-ttl-propagate disable [ forwarded | local ]
```

对于每个节点而言，TTL 复制会依据报文的来源进行区分操作，一种为数据报文来自外部节点，这种数据报文需要进行转发，包括收到 IP 报文后压入 MPLS 标签或者收到 MPLS 报文弹出标签还原为 IP 报文，将这一类输入数据定义为 forwarded 数据，其 TTL 复制可以通过"mpls ip-ttl-propagate disable forwarded"进行控制；另一种为报文来自节点本地，可以通过命令"mpls ip-ttl-propagate disable local"进行控制。

MPLS TTL 传播机制只对普通 IP/MPLS 数据转发过程起作用，对于 MPLS Echo 报文而言，其目的 IP 地址为环回接口地址，IP TTL 值总是设定为 1，TTL 传播机制在这种特殊报文上是不会生效的，因此无法通过 MPLS Echo 报文来对 TTL 传播机制进行实验。

可以通过 PC1 发送 ICMP 报文来验证 TTL 的传播机制。在各节点静态配置 5.5.5.5/32 的静态 MPLS 转发表项，确保 LSP 为节点 1→2→3→4→5，节点 4 作为 LSP 的倒数第二跳节点，开启倒数第二跳弹出功能；配置 PC1 接口网段 FEC

10.1.1.0/24 的静态路径为节点 5→4→3→2→1。节点 4 作为 LSP（FEC=5.5.5.5/32）的倒数第二跳节点，开启倒数第二跳弹出功能，在 PC1 上发送目的 5.5.5.5 的 IP ping 报文，IP 报文到达节点 1 后匹配 MPLS 转发表，压入 MPLS 标签并设定 TTL 值，MPLS 报文经过节点 2、3 的标签交换后到达节点 4，节点 4 弹出标签还原为 IP 报文并重新设定 IP 报文的 TTL 值后将 IP 报文转发给节点 5，从而到达目的节点。这里节点 1～节点 4 构成 MPLS 域，节点 1 为 MPLS 域入口节点，节点 4 为 MPLS 域出口节点。

在 PC1 上执行指令"ping 5.5.5.5 -T 100"，设置其发送的 IP 报文 TTL 值为 100。在节点 1 的 Gi0/0/0/2 接口上抓包，压入 MPLS 标签时从 IP 中使能 TTL 复制如图 3.12 所示，可以看到节点 1 在收到 IP 报文后首先将 IP TTL 值由 100 减为 99，然后查找本地 MPLS 转发表，命中 5.5.5.5 的 MPLS 转发表项，压入 MPLS 标签，标签值 222222，MPLS 的 TTL 值从 IP 报文中复制，所以也为 99。

```
11 152.4148…  10.1.1.10  5.5.5.5    ICMP     102 Echo (ping) request  id=0x8a0e, seq=2/512, ttl=99

Frame 11: 102 bytes on wire (816 bits), 102 bytes captured (816 bits) on interface -, id 0
Ethernet II, Src: 50:00:00:01:00:03 (50:00:00:01:00:03), Dst: 50:00:00:02:00:02 (50:00:00:02:00:02)
MultiProtocol Label Switching Header, Label: 222222, Exp: 0, S: 1, TTL: 99
Internet Protocol Version 4, Src: 10.1.1.10, Dst: 5.5.5.5
    0100 .... = Version: 4
    .... 0101 = Header Length: 20 bytes (5)
  > Differentiated Services Field: 0x00 (DSCP: CS0, ECN: Not-ECT)
    Total Length: 84
    Identification: 0x0e8a (3722)
  > Flags: 0x00
    Fragment Offset: 0
    Time to Live: 99
    Protocol: ICMP (1)
    Header Checksum: 0x340b [validation disabled]
    [Header checksum status: Unverified]
    Source Address: 10.1.1.10
    Destination Address: 5.5.5.5
Internet Control Message Protocol
```

图 3.12　压入 MPLS 标签时从 IP 中使能 TTL 复制

然后，在节点 1 上通过指令"mpls ip-ttl-propagate disable"关闭压入 MPLS 标签时的 TTL 复制操作，重复上述操作，在节点 1 的 Gi0/0/0/2 接口上抓包，压入 MPLS 标签时禁用 TTL 复制如图 3.13 所示，可以看到节点 1 将 IP 报文的 TTL 值由 100 减为 99 后，在压入 MPLS 标签时没有从 IP 报文复制 TTL 值，而是直接设置为缺省的 255。

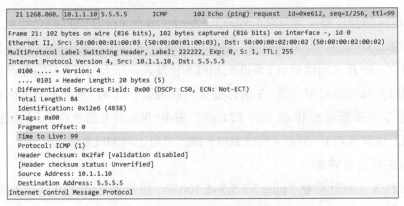

图 3.13　压入 MPLS 标签时禁用 TTL 复制

与此同时，在节点 4 的 Gi0/0/0/5 接口上抓包，查看 TTL 值变化情况，弹出 MPLS 标签时使能 TTL 复制（IP TTL<MPLS TTL）如图 3.14 所示。节点 1 发出 TTL 值为 255 的 MPLS 报文经过节点 2、节点 3、节点 4 后，TTL 值变为 252。节点 4 将 MPLS 标签弹出后，将 MPLS 报文的 TTL 值 252 复制到 IP 的 TTL 域，所以，IP 报文的 TTL 值应为 252。但是从图 3.14 的抓包结果来看，节点 4 在弹出标签时并未将 MPLS 报文的 TTL 值 252 复制到 IP 报文中。这是因为 IP 报文的 TTL 值比 MPLS 报文的 TTL 值小，为了避免报文环路，在转发过程中 TTL 值应该不断变小，尤其在 MPLS 与 IP 报文的 TTL 相互复制过程中，不允许用值大的 TTL 覆盖值小的 TTL。因此，这里看不出来节点 4 执行了 TTL 复制操作。

图 3.14　弹出 MPLS 标签时使能 TTL 复制（IP TTL<MPLS TTL）

为了验证这一规则，可以在 PC1 上执行"ping 5.5.5.5 -T 255"，将初始 IP 报文的 TTL 值设置为 255，并保持节点 1 的 TTL 复制功能被禁用。因节点 1 禁用了 TTL 复制，所以它在压入 MPLS 标签时将 MPLS 报文的 TTL 值设置为 255，该 MPLS 报文经过节点 2、节点 3、节点 4 后，TTL 值变为 252。节点 4 在弹出 MPLS 时，将 TTL 值复制到 IP TTL 域，此时 IP 报文的 TTL 值为 254，要大于 MPLS 报文的 TTL 值，因此，允许重写 IP 报文的 TTL 值，弹出 MPLS 标签时使能 TTL 复制（IP TTL>MPLS TTL）如图 3.15 所示，IP 报文的 TTL 值被复制为 252。

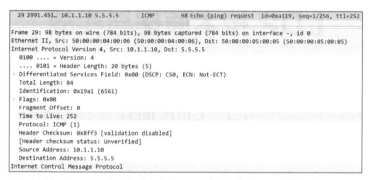

图 3.15　弹出 MPLS 标签时使能 TTL 复制（IP TTL>MPLS TTL）

现在，在节点 4 上通过指令"mpls ip-ttl-propagate disable"关闭 TTL 复制操作，PC1 的 ping 包设置 TTL=255，并保持节点 1 的 TTL 复制功能被禁用。在节点 4 的 Gi0/0/0/5 接口上抓包，弹出 MPLS 标签时禁用 TTL 复制如图 3.16 所示，此时节点 4 弹出 MPLS 报文时，不再将 MPLS 报文的 TTL 复制到 IP 报文中，因此，IP 报文的 TTL 值保持当时压入 MPLS 头部时的 254 不变。也就是说 IP 报文穿越 MPLS 域时，将 MPLS 域视为一个节点，TTL 值仅减 1。

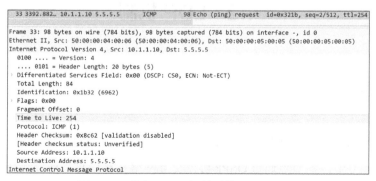

图 3.16　弹出 MPLS 标签时禁用 TTL 复制

第 4 章

LDP 技术原理与实战

4.1 LDP 简介

标签分发协议（LDP）负责为 FEC 分发标签，但自身无法独立运行，需要依赖 IGP。通过 LDP，路由器可能从任何一个邻居收到关于同一个 FEC 的标签信息，但是针对同一个 FEC 只会下发一条 MPLS 转发表项，路由器要依据 IGP 计算的最优下一跳来决定安装哪条 MPLS 转发表项。

LDP 采用多播消息进行邻居发现，Hello 消息通过 UDP 封装，使用 646 端口。通过 Hello 消息发现邻居后，节点之间建立 TCP 通道来传送标签分发消息并保证可靠传输，使用 TCP 646 端口。

4.2 LDP 机制

4.2.1 实验环境

LDP MPLS 实验拓扑如图 4.1 所示，所有路由器运行了 OSPF 协议，并处于同一个区域 0 中，各节点通过 OSPF 协议交互网络拓扑信息，建立到所有接口网段和主机地址的路由。

第 4 章　LDP 技术原理与实战

虚线框内的路由器接口上运行了 LDP，并假定节点 3 和节点 5 之间的链路不支持 MPLS，因此，节点 3 的 Gi0/0/0/5 接口和节点 5 的 Gi0/0/0/3 接口上不启动 LDP。

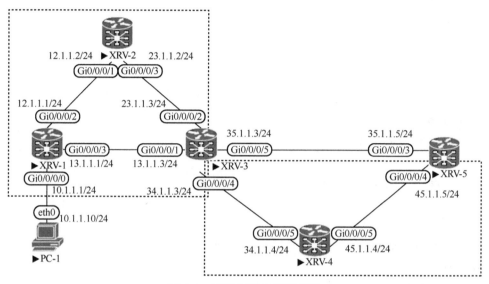

图 4.1　LDP MPLS 实验拓扑

4.2.2　协议配置

按照实验要求，在每个节点配置 OSPF 协议与 LDP，在各互联接口、本地环回接口上启动 OSPF 与 LDP，所有接口的 OSPF 链路代价均为 1。

LDP 在为本地路由分配标签时，要从本地标签池中申请标签，为了简化，节点 N 的本地标签池为 $N00000 \sim N99999$。各节点协议配置信息如下：

```
XRV-1:
router ospf 1
 area 0
  interface Loopback0
  !
  interface GigabitEthernet0/0/0/0
   passive enable
```

```
 !
 interface GigabitEthernet0/0/0/2
 !
 interface GigabitEthernet0/0/0/3
 !
!
mpls oam
!
mpls ldp
 interface GigabitEthernet0/0/0/2
 !
 interface GigabitEthernet0/0/0/3
 !
!
mpls label range table 0 100000 199999
```
XRV-2：
```
router ospf 1
 area 0
  interface Loopback0
  !
  interface GigabitEthernet0/0/0/1
  !
  interface GigabitEthernet0/0/0/3
  !
 !
!
mpls oam
!
mpls ldp
 interface GigabitEthernet0/0/0/1
 !
 interface GigabitEthernet0/0/0/3
 !
```

```
!
mpls label range table 0 200000 299999
```
XRV-3:
```
router ospf 1
 area 0
  interface Loopback0
  !
  interface GigabitEthernet0/0/0/1
  !
  interface GigabitEthernet0/0/0/2
  !
  interface GigabitEthernet0/0/0/4
  !
  interface GigabitEthernet0/0/0/5
  !
 !
!
mpls oam
!
mpls ldp
 address-family ipv4
 !
 interface GigabitEthernet0/0/0/1
 !
 interface GigabitEthernet0/0/0/2
 !
 interface GigabitEthernet0/0/0/4
 !
 interface GigabitEthernet0/0/0/5
!
mpls label range table 0 300000 399999
```
XRV-4:
```
router ospf 1
 area 0
```

```
   interface Loopback0
   !
   interface GigabitEthernet0/0/0/3
   !
   interface GigabitEthernet0/0/0/5
   !
  !
 !
mpls oam
!
mpls ldp
 interface GigabitEthernet0/0/0/3
 !
 interface GigabitEthernet0/0/0/5
 !
!
mpls label range table 0 400000 499999
```

XRV-5:

```
router ospf 1
 area 0
   interface Loopback0
   !
   interface GigabitEthernet0/0/0/3
   !
   interface GigabitEthernet0/0/0/4
   !
  !
 !
mpls oam
!
mpls ldp
 address-family ipv4
   label
     local
```

```
     advertise
      explicit-null
     !
    !
   !
  !
  interface GigabitEthernet0/0/0/4
  !
 !
mpls label range table 0 500000 599999
```

在完成上述配置后,可以查看各个节点的路由表,例如,对节点 1 上的本地路由、直连路由以及其他路由进行分类查看,可以看到,除了本地路由,节点 1 已经通过 OSPF 建立了到达所有其他主机和网段的路由。

```
RP/0/0/CPU0:XRV-1#show route ipv4 | include "L"
L    1.1.1.1/32 is directly connected, 02:59:29, Loopback0
L    10.1.1.1/32 is directly connected, 02:59:29, GigabitEthernet0/0/0/0
L    12.1.1.1/32 is directly connected, 02:59:29, GigabitEthernet0/0/0/2
L    13.1.1.1/32 is directly connected, 02:59:29, GigabitEthernet0/0/0/3
L    127.0.0.0/8 [0/0] via 0.0.0.0, 00:45:06
RP/0/0/CPU0:XRV-1#show route ipv4 | include "C"
C    10.1.1.0/24 is directly connected, 02:59:35, GigabitEthernet0/0/0/0
C    12.1.1.0/24 is directly connected, 02:59:35, GigabitEthernet0/0/0/2
C    13.1.1.0/24 is directly connected, 02:59:35, GigabitEthernet0/0/0/3
RP/0/0/CPU0:XRV-1#show route ipv4 | exclude "C" | exclude "L"
O    2.2.2.2/32 [110/2] via 12.1.1.2, 02:52:48, GigabitEthernet0/0/0/2
O    3.3.3.3/32 [110/2] via 13.1.1.3, 02:53:17, GigabitEthernet0/0/0/3
O    4.4.4.4/32 [110/3] via 13.1.1.3, 02:52:38, GigabitEthernet0/0/0/3
O    5.5.5.5/32 [110/3] via 13.1.1.3, 02:53:06, GigabitEthernet0/0/0/3
O    23.1.1.0/24 [110/2] via 12.1.1.2, 02:52:11, GigabitEthernet0/0/0/2
                 [110/2] via 13.1.1.3, 02:52:11, GigabitEthernet0/0/0/3
O    34.1.1.0/24 [110/2] via 13.1.1.3, 02:52:39, GigabitEthernet0/0/0/3
O    35.1.1.0/24 [110/2] via 13.1.1.3, 02:53:11, GigabitEthernet0/0/0/3
O    45.1.1.0/24 [110/3] via 13.1.1.3, 02:53:06, GigabitEthernet0/0/0/3
```

在建立起节点之间的路由之后，LDP 就可以为对应路由分配标签并相互分发，进而建立 MPLS 转发表项。上述实验使用 OSPF 建立路由，当然还可以使用其他 IGP 实现路由的建立。需要注意的是，不能基于 BGP 分发的路由来进行 LDP 标签分发，LDP 是不会为 BGP 路由分配标签的。

4.2.3　LDP 交互

1. LDP 邻居发现与会话建立

LDP 邻居可以分为直连邻居和非直连邻居。在接口上启动 LDP 后，LDP 将向 224.0.0.2 发送 UDP 封装的 LDP Hello 消息来发现直连邻居。对于非直连邻居无法通过多播消息自动发现，而只能使用单播发送 UDP 封装的 LDP Hello 消息，这个时候的远端 LDP 邻居只能由管理员手工配置。

在 LDP 发现邻居后，通过 TCP 承载的 LDP 消息进行邻居会话建立。按照邻居类型的不同，分为直连邻居会话和目标邻居会话，后者通常称为 Targeted LDP。

在节点 1 的 Gi0/0/0/3 接口上抓取的 LDP 直连邻居发现以及会话建立过程如图 4.2 所示。节点 1 的 Gi0/0/0/3 接口和节点 3 的 Gi0/0/0/1 接口上启动了 LDP，所以它们在该链路上通过多播进行 LDP 邻居发现（图 4.2 中第 10 号分组），并在 Hello 消息中携带可选字段 IPv4 Transport Address（IPv4 传输地址），以指示后续在 TCP 通道建立 LDP 会话时使用的 IP 地址，通常情况下这个地址为 LSR 的 Router ID。如同路由协议一样，在没有配置的情况下，路由器默认选取本地 Loopback 地址作为 LSR ID。

接下来，LDP 使用对方 Hello 消息中携带的传输地址进行 TCP 会话建立。因为通过 TCP 进行 LDP 消息交互，所以两个节点之间只需要一个 TCP 通道即可。LSR ID 大的节点（节点 3）作为主动发起方，与 LSR ID 小的邻居（节点 1）建立 TCP 会话（图 4.2 中第 44/45/46 号分组），之后，LSR 通过初始化消息协商版本号、标签分配方法（下游自主或按需）、定时器值等会话参数，如图 4.3 所示。

第 4 章　LDP 技术原理与实战

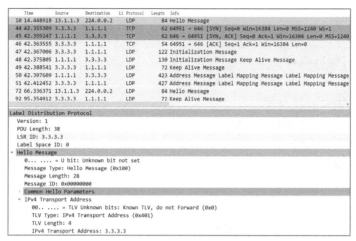

图 4.2　LDP 直连邻居发现以及会话建立过程

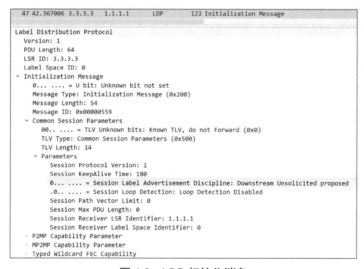

图 4.3　LDP 初始化消息

可以通过"show mpls ldp parameters"指令查看 LDP 会话参数，节点 1 的 LDP 参数信息如下：

```
RP/0/0/CPU0:XRV-1#show mpls ldp parameters
LDP Parameters:
  Role: Active
  Protocol Version: 1
```

```
Router ID: 1.1.1.1
Null Label:
  IPv4: Implicit
Session:
  Hold time: 180 sec
  Keepalive interval: 60 sec
  Backoff: Initial:15 sec, Maximum:120 sec
  Global MD5 password: Disabled
Discovery:
  Link Hellos:     Holdtime:15 sec, Interval:5 sec
  Targeted Hellos: Holdtime:90 sec, Interval:10 sec
  Quick-start: Enabled (by default)
  Transport address:
    IPv4: 1.1.1.1
Graceful Restart:
  Disabled
NSR: Disabled, Not Sync-ed
Timeouts:
  Housekeeping periodic timer: 10 sec
  Local binding: 300 sec
  Forwarding state in LSD: 15 sec
Delay in AF Binding Withdrawl from peer: 180 sec
Max:
  1500 interfaces (1200 attached, 300 TE tunnel), 2000 peers
OOR state
  Memory: Normal
```

邻居发现定时器的参数修改命令如下：

```
mpls ldp discovery (hello | targeted-hello) {holdtime | interval} <1-65536>
```

LDP 会话发现定时器的参数修改命令如下：

```
mpls ldp session holdtime <15-65535>
```

LDP 支持两种标签通告模式。一种为下游自主（Downstream Unsolicited，DU）分发模式，即 LSR 在发现有效 IGP 路由后，向所有邻居为这些路由分配对应的标签信息；另一种通过指令将标签通告模式修改为按需（Downstream on Demand，DoD）分发模式，即在接收到上游节点的请求后再进行标签分发，修改命令如下。
```
mpls ldp session downstream-on-demand with access-list
```
设置为 DoD 分发模式的 LDP 邻居 ID 需要被编辑在 acl 中，因此，在进行 DoD 分发模式修改前，必须先配置 acl。

2．地址消息通告

LDP 在交互完毕初始化消息后，需要进一步通告地址消息。因为采用 TCP 传输，所以地址消息、标签映射消息等都可以在一个 IP 分组中承载，如图 4.2 第 50/51 个分组所示。节点 1 向节点 3 通告的 LDP 地址消息如图 4.4 所示，可以看到节点 1 将所有本地接口地址都封装到该消息中进行了通告，而不管该接口是否启动了 LDP。接收节点从地址消息中提取邻居节点的接口信息，并建立"接口 IP 地址↔LSR ID"的映射关系，供后续计算 MPLS 转发表项时使用。

```
Time       Source    Destination  Li Protocol  Length  Info
50 42.397609 1.1.1.1  3.3.3.3        LDP         423    Address Message Label Mapping Message Label Mapping Message
Frame 50: 423 bytes on wire (3384 bits), 423 bytes captured (3384 bits) on interface -, id 0
Ethernet II, Src: 50:00:00:01:00:04 (50:00:00:01:00:04), Dst: 50:00:00:03:00:02 (50:00:00:03:00:02)
Internet Protocol Version 4, Src: 1.1.1.1, Dst: 3.3.3.3
Transmission Control Protocol, Src Port: 646, Dst Port: 64951, Seq: 77, Ack: 87, Len: 369
Label Distribution Protocol
  Version: 1
  PDU Length: 365
  LSR ID: 1.1.1.1
  Label Space ID: 0
  ▼ Address Message
     0... .... = U bit: Unknown bit not set
     Message Type: Address Message (0x300)
     Message Length: 26
     Message ID: 0x0000035b
   ▼ Address List
       00.. .... = TLV Unknown bits: Known TLV, do not Forward (0x0)
       TLV Type: Address List (0x101)
       TLV Length: 18
       Address Family: IPv4 (1)
     ▼ Addresses
         Address 1: 10.1.1.1
         Address 2: 12.1.1.1
         Address 3: 13.1.1.1
         Address 4: 1.1.1.1
  ▶ Label Mapping Message
```

图 4.4　LDP 地址消息

完成邻居之间的上述消息交互后，通过指令"show mpls ldp neighbor"可以查看指定邻居的参数信息：
```
RP/0/0/CPU0:XRV-1#show mpls ldp neighbor 3.3.3.3
Peer LDP Identifier: 3.3.3.3:0
  TCP connection: 3.3.3.3:64951 - 1.1.1.1:646
  Graceful Restart: No
  Session Holdtime: 180 sec
```

```
State: Oper; Msgs sent/rcvd: 640/642; Downstream-Unsolicited
Up time: 09:08:44
LDP Discovery Sources:
  IPv4: (1)
    GigabitEthernet0/0/0/3
  IPv6: (0)
Addresses bound to this peer:
  IPv4: (5)
    3.3.3.3     13.1.1.3     23.1.1.3     34.1.1.3     35.1.1.3
  IPv6: (0)
```

分析上面的打印信息，邻居 LSR 通过非 646 端口与本地建立 TCP 会话，说明了邻居 LSR 为 LDP 会话的主动方，会话的保持时间为 180s，采用 DU 分发模式分发标签。本地通过 Gi0/0/0/3 接口发现邻居 LSR，邻居 LSR 拥有 5 个接口，接口地址已列出。

3. MPLS 标签分发

最后，LSR 为每一条路由分配标签，并通过标签映射消息通告出去，如图 4.2 第 50 个分组所示。标签映射消息如图 4.5 所示，给出了节点 1 向节点 3 通告的多个标签映射消息中的一个，这里 FEC 前缀为 5.5.5.5/32，为其分配的标签值为 0x186a2，十进制为 100002。

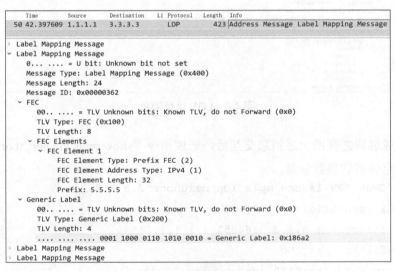

图 4.5　标签映射消息

通过指令"show mpls ldp bindings local"可以更为详尽地查看节点 1 为每一条路由分配的标签，具体如下：

```
RP/0/0/CPU0:XRV-1#show mpls ldp bindings local
Sat Feb  6 16:58:05.247 UTC


1.1.1.1/32, rev 22

     Local binding: label: ImpNull
2.2.2.2/32, rev 41
     Local binding: label: 100004
3.3.3.3/32, rev 37
     Local binding: label: 100000
4.4.4.4/32, rev 43
     Local binding: label: 100006
5.5.5.5/32, rev 39
     Local binding: label: 100002
10.1.1.0/24, rev 24
     Local binding: label: ImpNull
12.1.1.0/24, rev 26
     Local binding: label: ImpNull
13.1.1.0/24, rev 28
     Local binding: label: ImpNull
23.1.1.0/24, rev 44
     Local binding: label: 100007
34.1.1.0/24, rev 42
     Local binding: label: 100005
35.1.1.0/24, rev 38
     Local binding: label: 100001
45.1.1.0/24, rev 40
     Local binding: label: 100003
```

可以看到，在运行了 LDP 之后，节点 1 将自动为本地持有和通过 IGP 学习到的所有路由分配标签：为本地路由分配隐式空标签（ImpNull）；为非本地/直连路由从本地标签空间（100000～199999）中分配标签并分发。

LDP 在为路由分配标签的同时也会从邻居节点接收到对应的标签映射信息，例如，图 4.2 第 51 个分组为从节点 3 收到的标签映射消息。节点 1 除了从节点 3 收到标签映射消息，还会从节点 2 收到标签映射消息。每个 LSR 针对每一条前缀维护本地分配的标签和从每一个邻居收到的标签信息。例如，节点 1 针对这些路由维护的标签信息如下：

```
RP/0/0/CPU0:XRV-1#show mpls ldp bindings
1.1.1.1/32, rev 22
        Local binding: label: ImpNull
        Remote bindings: (2 peers)
            Peer                Label
            -----------------   ---------
            2.2.2.2:0           200000
            3.3.3.3:0           300000
2.2.2.2/32, rev 41
        Local binding: label: 100004
        Remote bindings: (2 peers)
            Peer                Label
            -----------------   ---------
            2.2.2.2:0           ImpNull
            3.3.3.3:0           300005
3.3.3.3/32, rev 37
        Local binding: label: 100000
        Remote bindings: (2 peers)
            Peer                Label
            -----------------   ---------
            2.2.2.2:0           200002
            3.3.3.3:0           ImpNull
4.4.4.4/32, rev 43
        Local binding: label: 100006
        Remote bindings: (2 peers)
            Peer                Label
            -----------------   ---------
            2.2.2.2:0           200003
```

```
            3.3.3.3:0               300003
5.5.5.5/32, rev 39
        Local binding: label: 100002
        Remote bindings: (2 peers)
            Peer                    Label
            ----------------        --------
            2.2.2.2:0               200004
            3.3.3.3:0               300002
10.1.1.0/24, rev 24
        Local binding: label: ImpNull
        Remote bindings: (2 peers)
            Peer                    Label
            ----------------        --------
            2.2.2.2:0               200001
            3.3.3.3:0               300001
12.1.1.0/24, rev 26
        Local binding: label: ImpNull
        Remote bindings: (2 peers)
            Peer                    Label
            ----------------        --------
            2.2.2.2:0               ImpNull
            3.3.3.3:0               300006
13.1.1.0/24, rev 28
        Local binding: label: ImpNull
        Remote bindings: (2 peers)
            Peer                    Label
            ----------------        --------
            2.2.2.2:0               200005
            3.3.3.3:0               ImpNull
23.1.1.0/24, rev 44
        Local binding: label: 100007
        Remote bindings: (2 peers)
            Peer                    Label
            ----------------        --------
```

```
                2.2.2.2:0              ImpNull
                3.3.3.3:0              ImpNull
34.1.1.0/24, rev 42
        Local binding: label: 100005
        Remote bindings: (2 peers)
                Peer                   Label
                -----------------      ---------
                2.2.2.2:0              200007
                3.3.3.3:0              ImpNull
35.1.1.0/24, rev 38
        Local binding: label: 100001
        Remote bindings: (2 peers)
                Peer                   Label
                -----------------      ---------
                2.2.2.2:0              200006
                3.3.3.3:0              ImpNull
45.1.1.0/24, rev 40
        Local binding: label: 100003
        Remote bindings: (2 peers)
                Peer                   Label
                -----------------      ---------
                2.2.2.2:0              200008
                3.3.3.3:0              300004
```

以 5.5.5.5/32 为例，可以看到，节点 1 为其分配了标签 100002，同时从 LDP 邻居节点 2 学习到其分配的标签 200004；从 LDP 邻居节点 3 学习到其分配的标签 300002。

4．MPLS 转发表项生成

节点在为 FEC 生成 MPLS 转发表项时，首先查找该 FEC 对应的路由，获得下一跳和出接口信息。然后依据下一跳 IP 地址，查找收到地址消息时建立的映射表，找到"接口 IP 地址 ↔ LSR ID"的映射关系，得知下一跳 LSR ID。在确保到达 LSR ID 的路由有效后，再以下一跳 LSR ID 和 FEC 前缀为索引，检索得到下一跳 LSR 为该 FEC 分配的标签,这就是出标签。在得到出标签和出接口信息后,结合本地为该 FEC 分配的标签,就可以生成标签交换表项。

在不指定 LSR ID 情况下,LDP 默认选用 Loopback 地址作为 LSR ID。而 LDP 会话的建立不依赖于 LSR ID,这就意味着在本地不存在到达 LSR ID 路由的情况下与之建立 LDP 会话。但是,LDP 在生成 MPLS 转发表项时,需要检查 LSR ID 的可达性,如果不可达,则会停止 MPLS 转发表项的计算,并认为下一跳不支持 LDP,将出标签置为"Unlabelled"。因此,在运行 LDP 的域内,IGP 必须将 Loopback 地址发布出去,以确保 LDP 能够正确生成 MPLS 转发表项。

以节点 1 的 FEC 5.5.5.5/32 的 MPLS 转发表建立过程为例,首先,本地为其分配入标签 100002,到达该前缀的出接口为 Gi0/0/0/3,下一跳 IP 为 13.1.1.3。依据该地址检索"接口 IP 地址↔LSR ID"的映射关系,得到下一跳 LSR ID 为 3.3.3.3,其为 5.5.5.5/32 分配的标签为 300002。于是,节点 1 为 5.5.5.5/32 建立的 MPLS 转发表项的入标签为 100002,出标签为 300002,出接口为 Gi0/0/0/3,标签操作为 SWAP。通过"show mpls forwarding"指令可以查看生成的 MPLS 转发表项信息,节点 1 建立的 MPLS 转发表和 5.5.5.5/32 的详细 MPLS 转发表信息如下:

```
RP/0/0/CPU0:XRV-1#show mpls forwarding
Local    Outgoing    Prefix        Outgoing       Next Hop      Bytes
Label    Label       or ID         Interface                    Switched
------   --------    ----------    -----------    ---------     --------
100000   Pop         3.3.3.3/32    Gi0/0/0/3      13.1.1.3      79536
100001   Pop         35.1.1.0/24   Gi0/0/0/3      13.1.1.3      0
100002   300002      5.5.5.5/32    Gi0/0/0/3      13.1.1.3      76052
100003   300004      45.1.1.0/24   Gi0/0/0/3      13.1.1.3      0
100004   Pop         2.2.2.2/32    Gi0/0/0/2      12.1.1.2      79572
100005   Pop         34.1.1.0/24   Gi0/0/0/3      13.1.1.3      0
100006   300003      4.4.4.4/32    Gi0/0/0/3      13.1.1.3      0
100007   Pop         23.1.1.0/24   Gi0/0/0/2      12.1.1.2      0
         Pop         23.1.1.0/24   Gi0/0/0/3      13.1.1.3      0
RP/0/0/CPU0:XRV-1#show mpls forwarding prefix 5.5.5.5/32 detail
Local    Outgoing    Prefix        Outgoing       Next Hop      Bytes
Label    Label       or ID         Interface                    Switched
------   --------    ----------    -----------    ---------     --------
```

```
100002      300002       5.5.5.5/32        Gi0/0/0/3        3.1.1.3          76052
    Updated: Feb  6 20:03:39.424
    Version: 60, Priority: 3
    Label Stack (Top -> Bottom): { 300002 }
    NHID: 0x0, Encap-ID: N/A, Path idx: 0, Backup path idx: 0,
Weight: 0
    MAC/Encaps: 14/18, MTU: 1500
    Outgoing Interface: GigabitEthernet0/0/0/3 (ifhandle 0x000000a0)
    Packets Switched: 1419
```

5. 显式空标签

当前 LDP 的实现是为本地路由分配隐式空标签（ImpNull），可以通过指令配置成显式空标签，命令如下：

```
mpls ldp address-family ipv4 label local advertise explicit-null
[for acl | to acl]
```

上述命令可以为所有本地路由通告显式空标签，也可以通过 acl 规则，只为特定本地路由通告显式空标签，或者只向特定 LSR 通告显式空标签。

当前实验中节点 5 就通过指令改变了系统默认行为，使得节点 5 为本地路由生成并通告显式空标签。在节点 4 上查看节点 5 为其本地路由 5.5.5.5/32 分配的标签信息，具体如下：

```
RP/0/0/CPU0:XRV-4#show mpls forwarding prefix 5.5.5.5/32 detail
Local    Outgoing    Prefix        Outgoing      Next Hop      Bytes
Label    Label       or ID         Interface                   Switched
------   ---------   -----------   -----------   -----------   -------
400000   Exp-Null-v4 5.5.5.5/32    Gi0/0/0/5     45.1.1.5      117500
    Updated: Feb  6 15:57:32.286
    Version: 62, Priority: 3
    Label Stack (Top -> Bottom): { 0 }
    NHID: 0x0, Encap-ID: N/A, Path idx: 0, Backup path idx: 0,
Weight: 0
    MAC/Encaps: 14/18, MTU: 1500
    Outgoing Interface: GigabitEthernet0/0/0/5 (ifhandle 0x000000e0)
    Packets Switched: 2218
```

6. Hello 消息与 Keepalive 消息

图 4.2 给出的 LDP 会话建立后，LSR 之间在通过 UDP 交互 Hello 消息的同时，还要通过 TCP 交互 Keepalive 消息。这两个消息都是用于邻居会话保活的，但作用不同。

LDP Hello 消息通过多播方式在每个 LDP 接口上通告，主要作用是实现主动的邻居发现，自动发现邻居 LSR。如果两个 LSR 之间存在多条链路，那么在每一条链路上都会交互 LDP Hello 消息。

不管 LSR 之间存在多少条链路，在一对 LSR 之间只会建立一个 LDP 会话。LDP Keepalive 消息用于 LSR 之间的 LDP 会话保活。

当连接某个邻居 LSR 的 LDP 链路失效后，通过 UDP Hello 的定时机制可以检测并发现，如果这两个 LSR 之间仍然存在其他存活的 LDP 链路，那么 LDP 会话将继续保持。同样，在 LDP 会话建立之后，在两个 LSR 之间再增加 LDP 链路，那么通过 UDP Hello 将自动发现邻居，但不会因为这个新邻居的出现而建立新的 LDP 会话，也不需要再次进行标签映射的通告。概括起来，UDP Hello 是每条链路的保活消息，而 TCP Hello 则是每一对邻居会话之间的保活消息。

4.3 转发路径验证

4.3.1 连通性验证

通过 LDP 能够实现 MPLS 域内所有路由的标签分配与分发，可以使用 MPLS ping 验证 LSP 的连通性。

下面在节点 1 上通过 MPLS ping 指令测试与目的 5.5.5.5/32 的连通性，具体指令如下。

```
RP/0/0/CPU0:XRV-1#ping mpls ipv4 5.5.5.5/32 fec-type ldp source
1.1.1.1 verbose
```

```
Sending 5, 100-byte MPLS Echos to 5.5.5.5/32,
     timeout is 2 seconds, send interval is 0 msec:

Codes: '!' - success, 'Q' - request not sent, '.' - timeout,
  'L' - labeled output interface, 'B' - unlabeled output interface,
  'D' - DS Map mismatch, 'F' - no FEC mapping, 'f' - FEC mismatch,
  'M' - malformed request, 'm' - unsupported tlvs, 'N' - no rx label,
  'P' - no rx intf label prot, 'p' - premature termination of LSP,
  'R' - transit router, 'I' - unknown upstream index,
  'X' - unknown return code, 'x' - return code 0

Type escape sequence to abort.

B     size 100, reply addr 13.1.1.3, return code 9
B     size 100, reply addr 13.1.1.3, return code 9
B     size 100, reply addr 13.1.1.3, return code 9
B     size 100, reply addr 13.1.1.3, return code 9
B     size 100, reply addr 13.1.1.3, return code 9

Success rate is 0 percent (0/5)
```

可以发现并未 ping 通目的地址，但是由 13.1.1.3 响应了带有 B 标志、返回代码为 9 的结果。

为了进一步探究原因，在 MPLS ping 操作的同时在节点 1 的 Gi0/0/0/3 接口上抓包，节点 1 发出的 MPLS Echo Request 分组如图 4.6 所示，第 1108 个分组，节点 1 发出的 MPLS Echo Request 分组压入了 MPLS 标签，标签值为 300002。针对该请求的 MPLS Echo Reply 分组由 IP 为 13.1.1.3 的地址发出，这个地址是节点 3 的 Gi0/0/0/1 的接口地址，节点 3 回应的 MPLS Echo Reply 分组如图 4.7 所示，第 1109 个分组，其响应的返回码为 9，从抓包结果可以看到节点 1 上的 MPLS ping 打印相关信息与抓包分析结果一致。

```
     Time          Source    Destination   Li Protocol     Length  Info
1108 1262.049… 1.1.1.1   127.0.0.1        MPLS ECHO      114 MPLS Echo Request
1109 1262.051… 13.1.1.3  1.1.1.1          MPLS ECHO       90 MPLS Echo Reply

MultiProtocol Label Switching Header, Label: 300002, Exp: 0, S: 1, TTL: 255
Internet Protocol Version 4, Src: 1.1.1.1, Dst: 127.0.0.1
User Datagram Protocol, Src Port: 3503, Dst Port: 3503
Multiprotocol Label Switching Echo
  Version: 1
> Global Flags: 0x0000
  Message Type: MPLS Echo Request (1)
  Reply Mode: Reply via an IPv4/IPv6 UDP packet (2)
  Return Code: No return code (0)
  Return Subcode: 0
  Sender's Handle: 0x000008ce
  Sequence Number: 1
  Timestamp Sent: Feb  7, 2021 13:54:31.833029764 UTC
  Timestamp Received: (0)Jan  1, 1970 00:00:00.000000000 UTC
> Vendor Private
v Target FEC Stack
    Type: Target FEC Stack (1)
    Length: 12
  v FEC Element 1: LDP IPv4 prefix
      Type: LDP IPv4 prefix (1)
      Length: 5
      IPv4 Prefix: 5.5.5.5
      Prefix Length: 32
      Padding: 000000
```

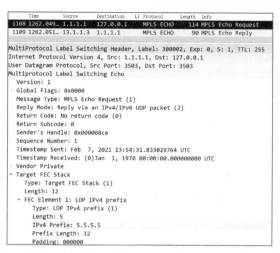

图 4.6　节点 1 发出的 MPLS Echo Request 分组

```
     Time          Source    Destination   Li Protocol     Length  Info
1108 1262.049… 1.1.1.1   127.0.0.1        MPLS ECHO      114 MPLS Echo Request
1109 1262.051… 13.1.1.3  1.1.1.1          MPLS ECHO       90 MPLS Echo Reply

Frame 1109: 90 bytes on wire (720 bits), 90 bytes captured (720 bits) on interface -, id 0
Ethernet II, Src: 50:00:00:03:00:02 (50:00:00:03:00:02), Dst: 50:00:00:01:00:04 (50:00:00:01:00:04)
Internet Protocol Version 4, Src: 13.1.1.3, Dst: 1.1.1.1
User Datagram Protocol, Src Port: 3503, Dst Port: 3503
Multiprotocol Label Switching Echo
  Version: 1
> Global Flags: 0x0000
  Message Type: MPLS Echo Reply (2)
  Reply Mode: Reply via an IPv4/IPv6 UDP packet (2)
  Return Code: Label switched but no MPLS forwarding at stack-depth RSC (9)
  Return Subcode: 1
  Sender's Handle: 0x000008ce
  Sequence Number: 1
  Timestamp Sent: Feb  7, 2021 13:54:31.833029764 UTC
  Timestamp Received: Feb  7, 2021 13:54:07.764678559 UTC
> Vendor Private
```

图 4.7　节点 3 回应的 MPLS Echo Reply 分组

因为 MPLS ping 的响应报文是由节点 3 回应的，进一步登录节点 3 查看其 MPLS 转发表项信息：

```
RP/0/0/CPU0:XRV-3#show mpls forwarding prefix 5.5.5.5/32 detail
Local    Outgoing     Prefix       Outgoing     Next Hop     Bytes
Label    Label        or ID        Interface                 Switched
-------- ------------ ------------ ------------ ------------ ----------
300002   Unlabelled   5.5.5.5/32   Gi0/0/0/5    35.1.1.5     7470
     Updated: Feb  7 13:46:26.396
     Version: 59, Priority: 3
```

```
    Label Stack (Top -> Bottom): { Unlabelled }
    NHID: 0x0, Encap-ID: N/A, Path idx: 0, Backup path idx: 0,
Weight: 0
    MAC/Encaps: 14/14, MTU: 1500
    Outgoing Interface: GigabitEthernet0/0/0/5 (ifhandle 0x000000e0)
    Packets Switched: 109
```

节点 3 针对 5.5.5.5/32 建立的 MPLS 转发表项为入标签 300002，出标签为 Unlabelled，这个不同于 Exp-Null-v4 或者 Imp-Null。该标志意味着此表项的出接口连接的下一跳节点不支持 MPLS，必须在该接口上将所有 MPLS 头剥离掉，还原为 IP 分组后才能转发。所以，节点 3 在收到 MPLS Echo Request 时，提取入标签 300002 后查表，命中上述表项，采取的操作就是剥离掉所有标签还原为 IP 分组后查找 IP 转发表。因为弹出 MPLS 标签后，IP 报文的目的 IP 地址为 127.0.0.1，所以该报文交付给节点 3 的本地协议栈进行处理，由 MPLS OAM 进程做出 MPLS Echo Reply 响应。节点 3 从报文中携带的 Target FEC 中发现本节点并非此报文请求的目的地，于是回应代码设定为 9。上述操作回答了为何中间节点 3 对 MPLS Echo Request 进行响应这一问题。

从上述过程中可以看出，发向节点 5 的 MPLS ping 包总是在节点 3 被处理，并由节点 3 进行响应，可以进一步思考并实验：如果发向节点 5 的不是 MPLS ping 包，而是以节点 5 为目的 IP 地址的普通 MPLS 报文，那么节点 3 是否也会对该报文进行响应。读者可以在 PC1 上发起 ping 5.5.5.5 操作并进行实验抓包分析。

还有一个问题值得探讨，那就是节点针对 5.5.5.5/32 的 MPLS 转发表项生成问题。节点 3 通过 LDP 为该前缀交互的标签信息如下：

```
RP/0/0/CPU0:XRV-3#show mpls ldp bindings 5.5.5.5/32
5.5.5.5/32, rev 43
    Local binding: label: 300002
    Remote bindings: (3 peers)
        Peer                  Label
        -----------------     --------
        1.1.1.1:0             100002
        2.2.2.2:0             200004
        4.4.4.4:0             400000
```

可以看到节点3还从节点1、节点2、节点4收到了它们为该FEC分配的标签信息，尤其值得关注的是节点4，它是节点3去往节点5的另一条路径上的节点。如果节点3将该FEC的入标签与节点4通告的标签进行关联，那么去往节点5的整条路径都可以支持MPLS转发了。那么为何节点3没有这么做？

回答这个问题，需要理解节点的MPLS转发表项生成原理，实际上在前一节已经进行了阐述，这里结合实例再次解析一下。

节点3先要甄别出本地到5.5.5.5/32的最优路由：下一跳为35.1.1.5，出接口为Gi0/0/0/5。依据下一跳地址35.1.1.5查找"接口IP地址↔LSR ID"映射表，发现由该地址检索不到任何LSR ID，因而无法基于下一跳地址关联到邻居LDP节点，也就是无法找到出标签。此时，认定去往该FEC的出接口为不支持MPLS的接口，针对该FEC的MPLS转发表项操作为Unlabelled。

4.3.2 标签交换路径验证

下面通过MPLS traceroute追踪节点1~节点5的LSP。在节点1上运行MPLS traceroute，得到结果如下：

```
RP/0/0/CPU0:XRV-1#traceroute mpls ipv4 5.5.5.5/32 fec-type ldp

  0 13.1.1.1 MRU 1500 [Labels: 300002 Exp: 0]
B 1 13.1.1.3 MRU 1500 [No Label] 0 ms
```

与MPLS ping类似，因为节点3去往节点5的最优路径上不支持MPLS，导致节点1与节点5的LSP断裂。

下面通过调整IGP链路代价的方式，让节点1去往节点5的最优路径均经过支持MPLS的节点，进而验证完整的LSP。具体操作就是：将节点1与节点3、节点3与节点5之间的OSPF链路代价设置为10000（其他链路代价默认为1），这样使得节点1去往节点5的最优IGP路径为节点1→2→3→4→5。因为LDP建立的LSP对IGP是完全依赖的，那么LSP也必然是节点1→2→3→4→5。

以节点1为例，修改OSPF链路代价的指令如下，节点3、节点5类比进行配置。

```
RP/0/0/CPU0:XRV-1(config)#router ospf 1 area 0 interface gi0/0/0/3
cost 10000
```

在节点 1 上通过 MPLS traceroute 验证节点 1 与节点 5 之间 LSP 经过的节点，得到如下结果：

```
RP/0/0/CPU0:XRV-1#traceroute mpls ipv4 5.5.5.5/32 fec-type ldp

  0 12.1.1.1 MRU 1500 [Labels: 200004 Exp: 0]
L 1 12.1.1.2 MRU 1500 [Labels: 300002 Exp: 0] 0 ms
L 2 23.1.1.3 MRU 1500 [Labels: 400000 Exp: 0] 0 ms
L 3 34.1.1.4 MRU 1500 [Labels: explicit-null Exp: 0] 10 ms
! 4 45.1.1.5 10 ms
```

可以看到去往 5.5.5.5/32 的 LSP 依次经过节点 1、2、3、4，到达节点 5。MPLS 分组沿着 LSP 转发过程中依次使用了标签 200004、300002、400000 和 explicit-null。

4.3.3　MPLS 分组的 TTL 超时响应

普通 MPLS 分组 TTL 超时时，检测到 TTL 超时的节点会产生 ICMP TTL 超时消息，下面对该 ICMP 消息的传送方式进行探讨分析。

在保持上述配置不变的情况下，在 PC1 上运行路径追踪指令，得到结果如下：

```
VPCS> trace 5.5.5.5 -P 1
trace to 5.5.5.5, 8 hops max (ICMP), press Ctrl+C to stop
 1   10.1.1.1    4.727 ms      3.020 ms     4.171 ms
 2   12.1.1.2   33.133 ms     25.077 ms    23.274 ms
 3   23.1.1.3   24.522 ms     22.697 ms    18.770 ms
 4   34.1.1.4   22.098 ms     20.300 ms    19.676 ms
 5   5.5.5.5    20.673 ms     16.532 ms    13.196 ms
```

可以看到 PC1 探测的去往节点 5 的路径与从节点 1 上进行路径追踪的结果一致。

在进行上述路径探测过程中，在节点 5 的 Gi0/0/0/4 接口上进行抓包，得到如图 4.8 所示的结果，节点 5 在该接口上收到了携带显式空标签的 MPLS 分组，这些 MPLS 分组承载了节点 2、节点 3、节点 4 发出的 ICMP TTL 超时报文。

第 4 章 LDP 技术原理与实战

```
Time        Source      Destination   Protocol  Length  Info
447 443.4174... 12.1.1.2    10.1.1.10    ICMP    186 Time-to-live exceeded (Time to live exceeded in transit)
448 443.4204... 12.1.1.2    10.1.1.10    ICMP    186 Time-to-live exceeded (Time to live exceeded in transit)
449 443.4431... 12.1.1.2    10.1.1.10    ICMP    186 Time-to-live exceeded (Time to live exceeded in transit)
450 443.4460... 12.1.1.2    10.1.1.10    ICMP    186 Time-to-live exceeded (Time to live exceeded in transit)
451 443.4676... 12.1.1.2    10.1.1.10    ICMP    186 Time-to-live exceeded (Time to live exceeded in transit)
452 443.4702... 12.1.1.2    10.1.1.10    ICMP    186 Time-to-live exceeded (Time to live exceeded in transit)
453 443.4942... 23.1.1.3    10.1.1.10    ICMP    186 Time-to-live exceeded (Time to live exceeded in transit)
454 443.4969... 23.1.1.3    10.1.1.10    ICMP    186 Time-to-live exceeded (Time to live exceeded in transit)
455 443.5185... 23.1.1.3    10.1.1.10    ICMP    186 Time-to-live exceeded (Time to live exceeded in transit)
456 443.5218... 23.1.1.3    10.1.1.10    ICMP    186 Time-to-live exceeded (Time to live exceeded in transit)
457 443.5384... 23.1.1.3    10.1.1.10    ICMP    186 Time-to-live exceeded (Time to live exceeded in transit)
458 443.5406... 23.1.1.3    10.1.1.10    ICMP    186 Time-to-live exceeded (Time to live exceeded in transit)
459 443.5600... 34.1.1.4    10.1.1.10    ICMP    186 Time-to-live exceeded (Time to live exceeded in transit)
460 443.5633... 34.1.1.4    10.1.1.10    ICMP    186 Time-to-live exceeded (Time to live exceeded in transit)
461 443.5835... 34.1.1.4    10.1.1.10    ICMP    186 Time-to-live exceeded (Time to live exceeded in transit)
462 443.5864... 34.1.1.4    10.1.1.10    ICMP    186 Time-to-live exceeded (Time to live exceeded in transit)
463 443.6053... 34.1.1.4    10.1.1.10    ICMP    186 Time-to-live exceeded (Time to live exceeded in transit)
464 443.6076... 34.1.1.4    10.1.1.10    ICMP    186 Time-to-live exceeded (Time to live exceeded in transit)
465 443.6273... 10.1.1.10   5.5.5.5      ICMP    110 Echo (ping) request  id=0x0f87, seq=0/0, ttl=4 (reply in 466)
466 443.6300... 5.5.5.5     10.1.1.10    ICMP    110 Echo (ping) reply    id=0x0f87, seq=0/0, ttl=255 (request in 465)
467 443.6461... 10.1.1.10   5.5.5.5      ICMP    110 Echo (ping) request  id=0x0f87, seq=0/0, ttl=4 (reply in 468)
468 443.6483... 5.5.5.5     10.1.1.10    ICMP    110 Echo (ping) reply    id=0x0f87, seq=0/0, ttl=255 (request in 467)
469 443.6636... 10.1.1.10   5.5.5.5      ICMP    110 Echo (ping) request  id=0x0f87, seq=0/0, ttl=4 (reply in 470)
470 443.6651... 5.5.5.5     10.1.1.10    ICMP    110 Echo (ping) reply    id=0x0f87, seq=0/0, ttl=255 (request in 469)
471 443.9956... 45.1.1.4    224.0.0.5    OSPF    106 Hello Packet
```

图 4.8 节点 5 的 Gi0/0/0/4 接口抓包

当 PC1 发出 TTL=2 的分组时，分组在节点 2 发生了 TTL 超时，于是节点 2 生成 ICMP TTL 超时报文，源 IP 为 12.1.1.2，目的 IP 为 PC1 地址 10.1.1.10，该 IP 报文并未进行查表，而是按照原来的 LSP，压入对应 MPLS 标签，使得该 ICMP TTL 超时报文被送达该 LSP 终点，由节点 5 进行标签弹出后，在节点 5 重新查表，并压入 10.1.1.10/24 对应的 MPLS 标签后返还给 PC1。与之类似，TTL=3 的 ICMP 分组在节点 3 超时后产生的 ICMP TTL 超时报文也是按照 LSP 到节点 5 后再环回去，在节点 5 上捕获的节点 3 发出的 ICMP TTL 超时报文如图 4.9 所示。该报文由节点 3 发出，经过节点 4 的标签交换后，沿着 LSP 到达节点 5，因此其 MPLS 标签为 Ipv 4 Explicit-Null。该 ICMP 报文承载了到达节点 3 时的原始 IP 报文以及 MPLS 头，通过该 ICMP 报文可以推断 PC1 发出的 ICMP 报文经历了什么：一方面，从 ICMP Multi-Part Extensions 中可以看到，到达节点 3 时的 MPLS 头部标签为 300002，TTL=1；另一方面，从原始 IP 报文可以看到到达节点 3 时的 IP 报文详情。实际过程是 PC1 发出 ICMP 报文，源 IP 和目的 IP 分别为 10.1.1.10 和 5.5.5.5，TTL 值设置为 3。该 ICMP 报文到达节点 1 时，将 TTL 值减为 2，并复制到新压入的 MPLS 头部 TTL 域，压入 MPLS 标签值 200004。然后沿着 LSP 经过节点 2，将标签交换为 300003，MPLS TTL 值由 2 变为 1，最终携带入标签为 300002、TTL=1 的 MPLS 报文到达节点 3。

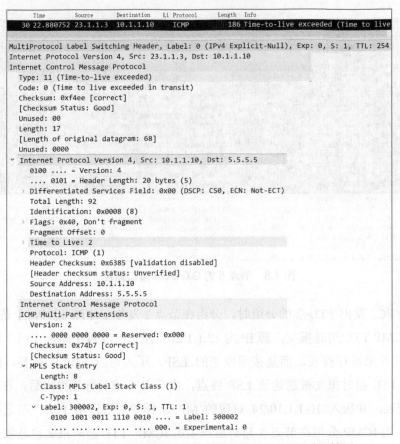

图 4.9 在节点 5 上捕获的节点 3 发出的 ICMP TTL 超时报文

对于节点 4 也是相同的情况,读者可以自行分析。所以,最终出现的现象就是 LSP 沿途节点产生的 TTL 超时报文并非由当前节点直接返回给源节点,而是先沿着 LSP 到达终点后再返回去。这就出现了在 LSP 终点接口上能够抓取到前序节点产生的 TTL 超时报文的情况。

MPLS TTL 超时分组沿着原先 LSP 送达终点后再重新查表路由回来,这是由 RFC3032 规定的。当然,可以通过配置指令"mpls ipv4 ttl-expiration-pop <n>"改变默认行为,由超时节点直接查表回应。也就是说如果 MPLS 标签数小于或等于 n,TTL 超时节点可以直接弹出标签,按照本地路由回送 TTL 超时分组,而不需要先送到 LSP 终点。

第 5 章

OSPF-SR 原理与实战

5.1 OSPF-SR 原理

5.1.1 基于 IGP 分发 SR 标签的优势

Segment Routing 数据面采用已有的 MPLS 或者 IPv6 技术。若采用 MPLS 作为数据面，则可重用 MPLS 标签承载 Segment ID 信息。

SR 具备任意路径编排能力，其关键支撑技术就是精细的路径元素定义能力。将通信路径实例化到域内，路径元素包括路径上的节点或节点组、节点之间的逻辑链路以及物理链路。任何一个路径元素都可以被分配一个 MPLS 标签。

已有的 LDP，只能为 FEC 分配标签，也就说标签分配的粒度限于前缀，且标签路径与对应前缀的最短路径相同，路径控制能力非常弱。同时，LDP 无法独立运行，其上下游标签之间的关联关系仍然要依赖 IGP。这些限制因素使得通过拓展 LDP 来支持 SR 的代价较高。

与之相比，IGP 通过交互链路状态数据实现拓扑构建，SR 所需的路径元素基本上可以映射到 IGP 的链路状态信息上。而且，IGP 自身具备路径计算功能，可以解决上下游标签的关联问题，直接在 IGP 进行 SR 标签分发成为首选。

5.1.2 MPLS 标签与 Segment ID

1．MPLS 标签的新内涵

SR 为了实现域内任意路径的编排，精细定义了路径元素。通过 IGP 为不同元素分配标签实现标签分发。通过组合这些标签，就可以实现对数据转发路径的任意控制。

概括起来 SR 为 5 类元素分配标签，路径元素示意图如图 5.1 所示。

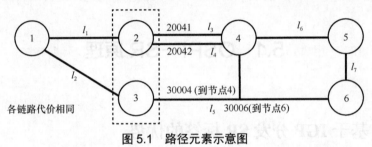

图 5.1　路径元素示意图

一是为节点分配标签。通过列出路径所经过的节点标签来描述一条路径，假定节点 N 的标签值为 16000+N，则{16001,16002,16004,16005}就描述了节点 1 依次经过节点 2、节点 4 到达节点 5 的路径。

二是为每条物理链路分配标签。当节点之间存在多条链路时，例如，节点 2 和节点 4 之间存在两条链路，要精确控制流量经过哪条链路，仅通过节点标签无法达成，此时就需要物理链路标签。例如，{16001,16002,20041,16004,16005} 表示该路径经过节点 2 和节点 4 之间的链路 l_3。

三是为每条逻辑链路分配标签。当多个节点之间共享一条物理链路时，仅通过物理链路无法甄别邻居之间的逻辑关系。例如，节点 3、节点 4、节点 6 位于同一个局域网（Local Area Network，LAN），虽然节点 3 通过一条链路连接到节点 4 和节点 6，但是存在节点 3 与节点 4、节点 3 与节点 6 以及节点 4 与节点 6 之间的逻辑连接，这些节点之间相当于存在 3 条逻辑链路，此时需要逻辑链路标签对这些链路进行区分。例如，{16001,16003,30004,16005}表示节点 3 经过节点 4 到达节点 5。

四是为逻辑节点分配标签。为了避免路径上单节点故障对数据转发的影响，可以将多个节点视为一个逻辑节点，并为之分配标签。逻辑节点中的某个实体节点失效将不会对转发路径产生影响。例如，将节点2和节点3视为一个逻辑节点 23 并分配标签 16023，则{16001,16023,16006}表示从节点 1 去往节点 6 的路径经过了逻辑节点 23，在缺省情况下，该逻辑路径对应路径为节点 1→3→6；当节点 3 失效时，该逻辑路径对应路径为节点 1→2→4→6。

五是为前缀分配标签。前缀标签并不是路径元素，为前缀分配标签并不是为了支撑 SR 的任意路径编排。实际上，它与 LDP 的标签分配一样，就是将标签与前缀建立映射关系，通过前缀能够检索到标签，将 IP 转发整合到 MPLS 转发上，便于统一数据面转发操作。

2．Segment ID 对 MPLS 标签内涵的拓展

SR 将路径分解为若干个 Segment，并为每个 Segment 分配一个 ID，即 SID。如果说 SR 为 MPLS 标签赋予了新内涵，让 MPLS 标签表示路径元素；那么 SID 则是 SR 对 MPLS 标签用法的定义和含义的拓展，让 MPLS 标签表达一段路径，而不再是路径元素。

节点标签唯一标识了一个节点，该标签的 SR 用法就是"去往该节点的等代价最短路径"。因此，SID 将节点标签与去往该节点的路径段关联，这个 SID 称为 NodeSID。

虚拟节点标签唯一标识了一组节点集合，该标签的 SR 用法就是"去往该节点组任意节点的等代价最短路径"。因此，SID 将虚拟节点标签与去往该节点组的路径段关联，这个 SID 称为 AnycastSID。

前缀标签唯一标识了此前缀，该标签的 SR 用法就是"去往该前缀的等代价最短路径"。因此，SID 将前缀标签与去往该节点的路径段关联，这个 SID 称为 PrefixSID。

物理链路标签唯一标识了一条链路，逻辑链路标签唯一标识了该链路上的一个节点，SID 的内涵与标签内涵一致，分别称为 AdjSID 和 LAN-AdjSID。

5.1.3　OSPF 链路状态信息与 SID 的关联

OSPF 链路状态信息中LSA1～LSA7 是与拓扑计算和路由计算相关的信息，

受报文格式定义约束，无法在其上拓展新字段承载 SR 相关信息。因此，SR 采用可扩展的不透明 LSA9～LSA11 来进行 SR 相关信息封装。

涉及了 5 种路径元素，包括实体节点、虚拟节点、前缀、物理链路和逻辑链路。可将这 5 种路径元素概括为两类：一类为前缀标识方式，包括实体节点、虚拟节点以及前缀，它们可统一采用"前缀+掩码"表示；另一类为链路标识方式，包括物理链路和逻辑链路。SID 的分配机制就是将 MPLS 标签绑定到相应前缀或者链路上。

OSPF 对前缀的描述方式为"路由类型+前缀+前缀长度"，对链路的描述方式为"链路类型+链路 ID+链路 Data"。因此，为了支持 SR 功能，拓展 OSPF 来通告 SID 信息的 LSA 格式可概括为如图 5.2 所示的消息结构，即在指定的前缀或链路描述信息后附加 SID 信息，从而实现为特定路径元素分配 SID 的目标。

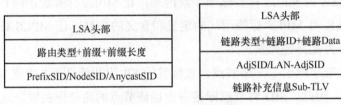

图 5.2　SR LSA 消息格式

5.1.4　SID 信息全域洪泛

OSPF 按照不透明 LSA 的洪泛范围进行链路状态信息扩散。SID 信息的传播范围与 OSPF 相应的链路状态信息传播范围保持一致。只有汇总前缀和外部前缀的 SID 才会跨越区域传送，其他类型的 PrefixSID 和所有的链路 SID 都只会在域内洪泛。

LDP 的标签传播方式与距离矢量协议有点类似，它不会对收到的标签映射信息进行中继，而是分配本地标签后向邻居通告。与之不同，OSPF 的 SID 信息遵从 OSPF 的消息传播机制，节点按照前述约束条件对收到的 SID（标签映射）消息进行洪泛。这种方法使得域内每个节点都持有相同的 SID 信息，这就意味着每个节点都掌握了相同的域内路径元素信息，为域内路径的任意编排提供了基础支撑。

5.1.5 MPLS 转发表项生成原理

1. NodeSID 的 MPLS 转发表项

由 SID 源发节点统一调控域内所有节点为该路径元素分配 MPLS 标签值。SR 节点通过计算，为非源发的 SID 生成本地 MPLS 标签。

每个 SR 节点单独保留一个标签空间，这个标签空间被称为 SR 全局块（Segment Routing Global Block，SRGB）。SID 源发节点不分配全局标签，而是分配全局标签索引，称之为 SID 索引。那么本地节点为相应路径元素分配的本地 MPLS 标签为本地 SRGB 起始值+SID 索引值。

在图 5.1 中，假定节点 N 的 SRGB 为$[N,N+6]$，每个节点为各自节点前缀分配 SID 索引为 N，例如，节点 5 的 SRGB 为 5～11，为自身节点分配 SID 索引为 5。通过 OSPF 的洪泛机制，域内任意节点都会建立如图 5.3 所示的标签索引表。

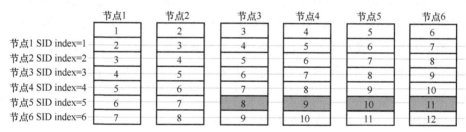

图 5.3 各节点的标签索引表

从节点 3 视角来看，它会收到节点 5 洪泛 SID 通告消息，该消息被节点 4 和节点 6 中继到节点 3。从该消息中，节点 3 解析得到节点 5 为节点前缀分配了 index 为 5 的 SID，通过叠加节点 5 的 SRGB，得知节点 5 使用 MPLS 标签 10 来标识节点 5。基于该 SID index，节点 3 通过叠加本地 SRGB，从而计算出本地为节点 5 前缀分配的 MPLS 标签为 8，这个标签是节点 3 依据节点 5 通告的 SID index 隐式计算出来的，这是节点 3 为节点 5 前缀分配的入标签，但是节点 3 不会通告这个标签。对于节点 3 而言，其关于节点 5 的出标签如何计算呢？这要依赖于 IGP 路由。由于节点 3 去往节点 5 前缀的最优下一跳为节点 4 和节点 6，它们是等代价路径，所以需要得到节点 4 和节点 6 的入标签。与节点 3 的入标签计算方式相同，节点 4 和节点 6 不会为节点 5 前缀显式分配并通告标签，节点 3 仍然通过在 SID index 上叠加节点 4、节点 6 的 SRGB，计算得到节点 4 和节点 6 为节点 5 前缀生成的标签：分别为 9 和 11。于是节点 3 为针对节点 5 前缀建立的 MPLS 转发表项为：

本地标签 Local	出标签 Out	下一跳 Nexthop	出接口 Oif
3	9	Node4	l_5
	11	Node6	l_5

SRGB 不同时关于节点 5 的 MPLS 转发表项如图 5.4 所示。首先，节点 5 在全域洪泛了自身与 SID 绑定信息，其他节点通过计算得出与之关联的本地标签，这些标签并不进行通告。每个非 SID 源发节点都采用计算方式得到该 SID 在本地的映射标签信息。其次，节点本地标签与出标签的关联关系依赖于 IGP 路由，将本地标签与最优下一跳标签进行关联绑定，形成 MPLS 转发表。需要说明的是 SR 支持以不同算法计算下一跳，目前规定算法有最短路径优先（SPF）算法和严格 SPF 算法。SID 源发节点在将 SID 绑定到前缀上进行通告时，SID 信息中携带计算关联标签时所采用的算法，其他节点都需要依据 SID 指定的算法来计算下一跳进而完成本地标签与下一跳标签的关联。最后，关于倒数第二跳节点的识别问题，当节点在进行本地标签与出标签的关联时，如果发现下一跳节点就是目的节点，那么出标签直接设置为显式空标签 0 或者隐式空标签（即 POP 操作）。当然，究竟采用哪种方式，要依据目的节点 5 的能力来定，所以，节点 5 在通告 SID 信息时会携带标志位，以指示倒数第二跳节点应该采用何种操作。

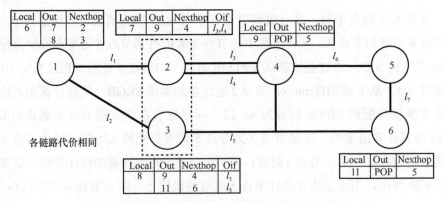

图 5.4　SRGB 不同时关于节点 5 的 MPLS 转发表项

第 5 章 OSPF-SR 原理与实战

通常情况下，所有节点的 SRGB 会设置为相同，为了阐述方便，假定下文中所有节点的 SRGB 为[16000，19999]，那么 SRGB 相同时关于节点 5 的 MPLS 转发表如图 5.5 所示。在这里我们可以进一步理解 MPLS 标签与 SID 内涵的不同：标签 16005 唯一标识了节点 5，NodeSID 16005 则表达了到达节点 5 的最短路径。携带 16005 标签的数据分组到达任何一个节点，都将沿着最短路径到达节点 5。图 5.5 中，节点 1 去往节点 5 存在 3 条 ECMP，分别为 P1：1→2→4→5；P2：1→3→4→5；P3：1→3→6→5。携带 16005 的 MPLS 分组到达节点 1 后，将会通过这 3 条路径以负载均衡方式分发，并最终达到节点 5。

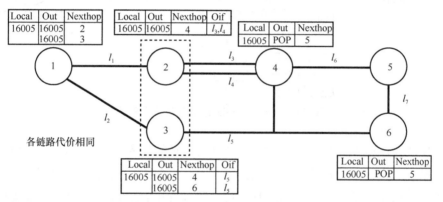

图 5.5　SRGB 相同时关于节点 5 的 MPLS 转发表项

2. AdjSID 的 MPLS 转发表项

对于链路标签或者 AdjSID 而言，二者含义一样，均用于标识链路，而不是路径，只有与之关联的节点才需要这些信息来确定从哪条链路上转发数据。因此，AdjSID 对应的标签不消耗 SRGB 空间。每个节点都会为本地链路分配标签，并以 AdjSID 形式与链路状态信息绑定后在全域洪泛。关于 AdjSID 的 MPLS 转发表项如图 5.6 所示，节点 2 和节点 3 为物理链路和逻辑链路分配的 AdjSID 标注于链路侧，节点 2 用 20001、20002、20003 标识物理链路 l_3、l_4 和 l_1，节点 3 用 20001、20002 标识物理链路 l_5 上去向节点 4、节点 6 的逻辑链路，用 20003 标识物理链路 l_2。

AdjSID 标签用于与本地链路建立关系，用于指示本地转发操作，它独立于其他节点，也不会与下游标签有任何关联，因此，标签转发表项只会与标签弹出操作、出接口、下一跳有关。节点 2 和节点 3 针对 AdjSID 建立的 MPLS 转发表项如图 5.6 所示，可以看到链路标签仅关联 POP 操作，而与其他节点标签无关。

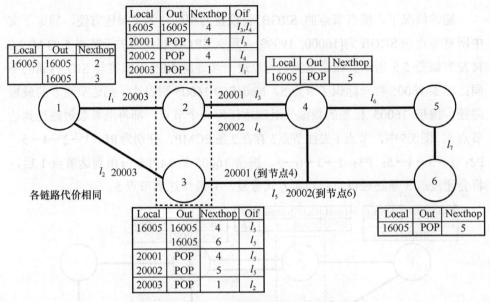

图 5.6　关于 AdjSID 的 MPLS 转发表项

NodeSID 和 AdjSID 的组合使用，将实现路径的任意编排。例如，分析节点 1 去往节点 5 的路径，如果期望路径经过节点 2、节点 4、节点 5 且经过链路 l_3，那么可以将送达节点 1 的 MPLS 标签栈设置为{16002, 20001, 16005}；如果期望路径经过节点 3、节点 6、节点 5，那么可以将送达节点 1 的 MPLS 标签栈设置为{16003, 20002, 16005}。数据分组在这两种转发需求下，节点 1 去往节点 5 的数据转发过程如图 5.7 所示。

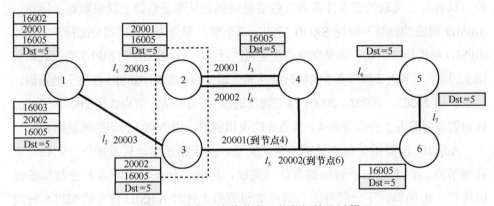

图 5.7　节点 1 去往节点 5 的数据转发过程

第 5 章 OSPF-SR 原理与实战

3. AnycastSID 的 MPLS 转发表项

SR 可以为不同节点分配相同的标签，到达该标签的最短路径标识就是 AnycastSID。AnycastSID 与 NodeSID 的 SID 分配原则、本地标签计算方式以及与出标签关联操作上是一致的。

AnycastSID 的 MPLS 转发表项如图 5.8 所示，所有节点的 SRGB 为[16000,19999]，将节点 2 和节点 3 视为一个逻辑节点 23，并配置相同 IP 地址，让节点 2 和节点 3 各自源发相同的 AnycastSID index 23。对于节点 6 而言，计算得到 AnycastSID 的本地标签为 16000+23=16023，本地到达源发节点 2 的最优下一跳为 4，路径代价为 2；到达源发节点 3 的最优下一跳为 3，路径代价为 1。因此，节点 6 将到达节点 3 的路径确定为到达逻辑节点 23 的最优路径，并使用到达节点 3 的标签作为出标签。于是得到 AnycastSID 对应的 MPLS 转发表项为将 16023 交换为 16023。又因为节点 6 是去往节点 3 的倒数第二跳，按照倒数第二跳的标签生成规则，16023 被交换为 0 或者弹出。如果节点 3 失效，使得节点 6 去往节点 3 的路径劣于去往节点 2 的路径，节点 6 会自动将出标签更换为下一跳节点 4 的标签 16023，并把下一跳设置为节点 4。若链路 l_5 失效，使得节点 6 使用节点 5 作为下一跳，则将出标签更换为下一跳节点 5 的标签 16023，并把下一跳设置为节点 5。

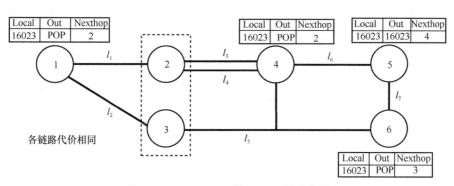

图 5.8 AnycastSID 的 MPLS 转发表项

AnycastSID 的运用将利于简化网络管理操作。在图 5.8 中，节点 2 和节点 3 为网络安全设备，节点 6 去往节点 1 方向的流量必须经过网络安全设备检查。在不使用 AnycastSID 的条件下，在节点 6 部署数据转发策略，去往目的节点 1 的分组压入标签 16002；当网络拓扑改变，使得节点 6 无法达到节点 2 时，需要将节点 6 的策略改变为去往目的节点 1 的分组压入标签 16003。这就意味着要依据网络拓扑改变动态，修改节点 6 的转发策略。当采用 AnycastSID 后，节点的策略配置为去往节点 1 的流量压入标签 16023。无论拓扑如何变化，都不需要更改节点 6 的转发策略。这是因为节点 6 会依据网络拓扑变化情况自适应地对标签进行交换，且必然经过节点 2 或节点 3，从而简化了管理难度。

5.2 域内 SR 实验

5.2.1 实验环境

域内 SR 实验拓扑如图 5.9 所示，所有路由器运行了 OSPF 协议，并处于同一个区域 0 中，节点 3、节点 4、节点 6 接入 LAN，其他节点之间为点到点链路类型，所有 OSPF 链路代价均为 1。在节点 N 设置环回接口，并配置 IP 地址 $N.N.N.N/32$。各节点通过 OSPF 协议交互网络拓扑信息，建立到所有接口网段和主机地址的路由。

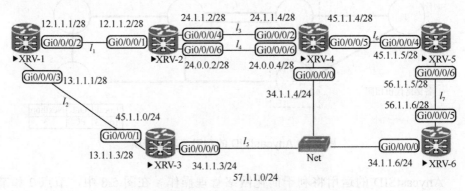

图 5.9 域内 SR 实验拓扑

5.2.2 协议配置

按照实验要求，为每个节点配置 OSPF 协议，在 OSPF 进程中启动 SR 功能，并手工指定节点的 SID index，节点 1 的协议配置信息如下：

```
router ospf 1
 segment routing mpls
 area 0
  interface Loopback0
   prefix-sid index 1
  !
  interface GigabitEthernet0/0/0/2
  !
  interface GigabitEthernet0/0/0/3
  !
 !
!
mpls oam
!
segment-routing
 global-block 16000 23999
!
```

其他节点的 OSPF 配置参照节点 1，并根据拓扑修改 area 和 SID index。

5.2.3 协议交互过程

1. SR 能力通告

OSPF 默认条件下不开启 SR 功能。SR 能力开启指令为：
```
router ospf <ID> segment-routing mpls
```

在开启 SR 功能后，OSPF 便通过不透明 LSA 进行能力通告，通过"路由器信息 LSA"通告当前节点支持的通用能力、SR 算法、SRGB，以及节点最大段深度（MSD）等。链路上抓取的节点 1 通告的 SR 能力如图 5.10 所示，从报文内容上看，节点 1 支持算法 0（SPF）和算法 1（严格 SPF）；SID/Label 起始值为 16000，深度为 8000，于是得到 SRGB 范围为[16000, 23999]；支持的最大标签栈深度为 10。

```
✓ LSA-type 10 (Opaque LSA, Area-local scope), len 60
    .000 0000 0000 0001 = LS Age (seconds): 1
    0... .... .... .... = Do Not Age Flag: 0
  > Options: 0x20, (DC) Demand Circuits
    LS Type: Opaque LSA, Area-local scope (10)
    Link State ID Opaque Type: Router Information (RI) (4)
    Link State ID Opaque ID: 0
    Advertising Router: 1.1.1.1
    Sequence Number: 0x80000002
    Checksum: 0xf05d
    Length: 60
  ✓ Opaque Router Information LSA
    ✓ Router Informational Capabilities
        TLV Type: Router Informational Capabilities (1)
        TLV Length: 4
      > RI Options: 0x60, (GRH) Graceful Restart Helper, Stub Router Support
    ✓ SR-Algorithm
        TLV Type: SR-Algorithm  (8)
        TLV Length: 2
        SR-Algorithm: Shortest Path First (0)
        SR-Algorithm: Strict Shortest Path First (1)
    ✓ SID/Label Range  (Range Size: 8000)
        TLV Type: SID/Label Range (9)
        TLV Length: 12
        Range Size: 8000
        Reserved: 00
      ✓ SID/Label Sub-TLV  (SID/Label: 16000)
          TLV Type: SID/Label (1)
          TLV Length: 3
          SID/Label: 16000
    ✓ Node MSD
        TLV Type: Node MSD (12)
        TLV Length: 2
        MSD Type: Base MPLS Imposition (1)
        MSD Value: 10
```

图 5.10 节点 1 通告的 SR 能力

默认采用 SRGB[16000, 23999]，修改指令为：
`router ospf <ID> segment-routing global-block <start> <end>`

2. AdjSID 通告与转发表的建立

在启动 SR 功能后，节点自动为所有链路进行 AdjSID 的分配和通告。节点 4 连接的链路类型比较丰富，既包括点到点链路又包括广播链路。AdjSID 信息采

第 5 章　OSPF-SR 原理与实战

用洪泛方式扩散，可以在任一点查看节点 4 通告的链路状态数据库中 AdjSID 信息，节点 6 记录的 AdjSID 信息如下：

```
RP/0/0/CPU0:XRV-6#show ospf database opaque-area adv-router 4.4.4.4
```
链路 l_5 的 AdjSID 信息
```
    Extended Link TLV: Length: 48
      Link-type : 2
      Link ID       : 34.1.1.6
      Link Data     : 34.1.1.4
      LAN Adj sub-TLV: Length: 11
        Flags       : 0x60
        MTID        : 0
        Weight      : 0
        Neighbor ID : 3.3.3.3
        Label       : 24000
      Adj sub-TLV: Length: 7
        Flags       : 0x60
        MTID        : 0
        Weight      : 0
        Label       : 24001
      Link MSD sub-TLV: Length: 2
        Type: 1, Value 10
```
链路 l_3 的 AdjSID 信息
```
    Extended Link TLV: Length: 56
      Link-type : 1
      Link ID       : 2.2.2.2
      Link Data     : 24.1.1.4
      Adj sub-TLV: Length: 7
        Flags       : 0x60
        MTID        : 0
        Weight      : 0
        Label       : 24002
      Local-ID Remote-ID Private sub-TLV: Length: 12
        Local Interface ID: 6
```

```
        Remote Interface ID: 8
    Remote If Address sub-TLV: Length: 4
        Neighbor Address: 24.1.1.2
    Link MSD sub-TLV: Length: 2
        Type: 1, Value 10
链路 $l_4$ 的 AdjSID 信息
    Extended Link TLV: Length: 56
      Link-type    : 1
      Link ID      : 2.2.2.2
      Link Data    : 24.0.0.4
      Adj sub-TLV: Length: 7
        Flags      : 0x60
        MTID       : 0
        Weight     : 0
        Label      : 24004
      Local-ID Remote-ID Private sub-TLV: Length: 12
        Local Interface ID: 10
        Remote Interface ID: 10
      Remote If Address sub-TLV: Length: 4
        Neighbor Address: 24.0.0.2
      Link MSD sub-TLV: Length: 2
        Type: 1, Value 10
链路 $l_6$ 的 AdjSID 信息
    Extended Link TLV: Length: 56
      Link-type    : 1
      Link ID      : 5.5.5.5
      Link Data    : 45.1.1.4
      Adj sub-TLV: Length: 7
        Flags      : 0x60
        MTID       : 0
        Weight     : 0
        Label      : 24003
      Local-ID Remote-ID Private sub-TLV: Length: 12
        Local Interface ID: 9
```

```
    Remote Interface ID: 7
Remote If Address sub-TLV: Length: 4
    Neighbor Address: 45.1.1.5
Link MSD sub-TLV: Length: 2
    Type: 1, Value 10
```

链路 l_5 为 LAN，Link-type=2，通过 34.1.1.4 接入该 LAN，该子网指定的路由器（DR）为 34.1.1.6（即节点 6），为该子网分配标签 24001，实际上就是节点 4 为节点 6 这个逻辑邻居分配的 AdjSID。因为该子网上还存在另外一个邻居节点 3，所以针对这个逻辑邻居 3.3.3.3 分配标签 24000，即 LAN-AdjSID。同时表明本地支持的链路最大标签栈深度为 10。

链路 l_3 为点到点类型，Link-type=1，节点通过本地链路接口（IP 为 24.1.1.4）连接邻居 2.2.2.2，并分配 AdjSID 为 24002。由于节点之间会存在多条链路，为了对这条链路做清晰的界定并与其他链路进行区分，进一步通过远端接口地址子 TLV（Remote If Address sub-TLV）给出当前链路连接到 2.2.2.2 的 24.1.1.2 所标识的接口上，甚至通过本地与远端接口 ID 子 TLV（Local-ID Remote-ID Private sub-TLV）来表明这条链路在两端的接口 ID 信息，本地接口索引为 6，对端接口索引为 8。可通过以下指令查看与核验接口索引信息：

```
RP/0/0/CPU0:XRV-4#show snmp interface Gi0/0/0/2 ifindex
ifName : GigabitEthernet0/0/0/2   ifIndex : 6
RP/0/0/CPU0:XRV-2#show snmp interface g0/0/0/4 ifindex
ifName : GigabitEthernet0/0/0/4   ifIndex : 8
```

链路 l_4 和 l_6 均为点到点类型，依据同样的原理，为其分配 AdjSID 24004 和 24003。

当节点为链路分发 AdjSID 信息后，在本地将产生对应的 MPLS 转发表项。由于 AdjSID 仅与本地链路有关，因此，所有标签对应的出标签都是 POP 并转发到指定链路的操作，节点 4 关于 AdjSID 的 MPLS 转发表项如下：

```
RP/0/0/CPU0:XRV-4#show mpls forwarding
Local    Outgoing     Prefix              Outgoing     Next Hop        Bytes
Label    Label        or ID               Interface                    Switched
------   ----------   -----------------   ----------   -----------     --------
24000    Pop          SR Adj (idx 0)      Gi0/0/0/0    34.1.1.3        0
```

```
24001      Pop         SR Adj (idx 0)    Gi0/0/0/0    34.1.1.6    0
24002      Pop         SR Adj (idx 0)    Gi0/0/0/2    24.1.1.2    0
24003      Pop         SR Adj (idx 0)    Gi0/0/0/5    45.1.1.5    0
24004      Pop         SR Adj (idx 0)    Gi0/0/0/6    24.0.0.2    0
```

3. NodeSID 通告与转发表的建立

为了对全局 SID 进行统一管理，OSPF 不会自动为节点分配 SID，只能由管理员手工配置。为节点或者前缀分配 SID 的指令为：

```
router ospf <ID> area X interface <ifname>
prefix-sid [strict-spf] (absolute | index) <value> [explicit-null]
[n-flag-clear]
```

SID 配置指令在接口模式下，配置的 SID 直接与接口上的 IP 地址关联。SID 配置可以采用绝对值模式或者索引模式，前者是直接配置标签值，后者则是配置索引，无论采用哪种模式，在拓展前缀 TLV 的 Prefix SID Sub-TLV 中携带的均为索引值。

参数 explicit-null 用于禁止倒数第二跳弹出功能，即倒数第二跳节点在进行本地标签与出标签关联绑定时，必须把出标签设置为显式空标签 0。Prefix SID Sub-TLV 的 Flags 中的 NP 和 E 标志位用以承载该参数。如果设置了 explicit-null，则 NP=E=1，否则 NP=E=0。

参数 n-flag-clear 是 PrefixSID 和 NodeSID 的区分标志，用于告知其他节点这个 SID 是绑定到 Prefix 上还是 Node 上，它在拓展前缀 TLV 的 Flags 字段携带，如果配置了 n-flag-clear，则 Node Flag 设置为 1。

参数 strict-spf 用于告知其他节点在计算到达该 SID 的路径时是采用哪种 SPF 算法，节点在能力通告时会通告自己支持 SPF 算法种类，在为前缀分配 SID 信息时，需要管理员手工指定算法类型（默认采用算法 0），该字段在 Prefix SID Sub-TLV 的 SR-Algorithm 字段中携带。

例如，对节点 5 进行了如下配置：

```
router ospf 1 area 0 interface loopback0
    prefix-sid absolute 16005 explicit-null
    prefix-sid strict-spf index 55
```

拓展前缀 TLV 示例如图 5.11 所示，给出了节点 5 通告的 NodeSID 信息。从拓展前缀 TLV 的 Flags 字段的 Node 标志可知，该通告为 NodeSID 通告，Node

前缀信息为 5.5.5.5/32。从 PrefixSID Sub-TLV 中可知，为该前缀分配了两个 SID，一个采用配置 SID 值方式，其使用算法 0 进行最短路径计算，并要求倒数第二跳节点采用显式空标签；另一个采用配置索引方式，采用算法 1 进行最短路径计算，要求倒数第二跳节点采用隐式空标签。无论配置采用的是 SID 还是索引，在消息通告时，统一通告索引。

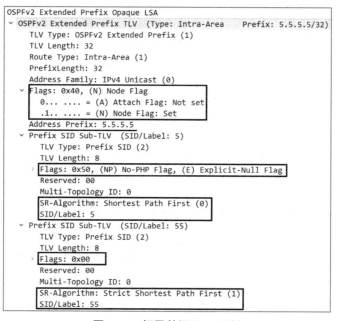

图 5.11　拓展前缀 TLV 示例

任意节点收到 NodeSID 信息后，将基于 SRGB+index 机制计算并生成本地 MPLS 转发表项。以节点 4 为例，其针对节点 5 通告的两个 SID 信息，分别计算转发表项。倒数第二跳节点的识别规则是下一跳节点 ID 是通告该 SID 的节点 ID。如果发现自己为倒数第二跳节点，还需要进一步查看源发节点的 NP、E 标志，进而按照源发节点需求的方式进行标签交换。因为节点 4 为倒数第二跳节点，且 SID=5 携带了 NP=1、E=1 标志，所以针对 16005 的出标签为显式空标签，针对 16055 的出标签为隐式空标签。因此，节点 4 的 MPLS 转发表项为：

```
RP/0/0/CPU0:XRV-4#show mpls forwarding
Local    Outgoing    Prefix       Outgoing      Next Hop    Bytes
Label    Label       or ID        Interface                 Switched
```

```
------  ------------   ------------------   ----------   ---------   --------
16005   Exp-Null-v4    SR Pfx (idx 5)       Gi0/0/0/5    45.1.1.5    0
16055   Pop            SR Pfx (idx 55)      Gi0/0/0/5    45.1.1.5    0
```

4. AnycastSID 通告与转发表的建立

AnycastSID 的配置模式与 NodeSID 一致，但要配置"n-flag-clear"参数，以便在拓展前缀 TLV 的 Flag 中将 Node Flag 设置为 0，表示这并非节点 SID。

为了进行 AnycastSID 通告，需要在相应节点配置一个 Anycast 地址。在节点 2 与节点 3 上配置 Anycast 地址为 23.1.1.23/32，如下：

```
RP/0/0/CPU0:XRV-2#interface loopback23 ipv4 address 23.1.1.23 255.2
55.255.255
RP/0/0/CPU0:XRV-3#interface loopback23 ipv4 address 23.1.1.23 255.2
55.255.255
```

进入 OSPF 进程配置 AnycastSID，如下：

```
RP/0/0/CPU0:XRV-2#router ospf 1 area 0 interface loopback23 prefix-
sid index 23 n-flag-clear
RP/0/0/CPU0:XRV-3#router ospf 1 area 0 interface loopback23 prefix-
sid index 23 n-flag-clear
```

通过上述配置，节点 2 和节点 3 的环回接口上配置了相同的 IP 地址，并分别为该地址分配标签 16023。网络上其他节点基于各自到节点 2 和节点 3 的最优路径计算下一跳，进而生成关于 AnycastSID 的 MPLS 转发表项。

对于节点 6 而言，它到 Anycast 节点 2 的路径代价为 2，到节点 3 的代价为 1，因此，最优下一跳为节点 3，又因为是倒数第二跳，因此，MPLS 转发表项为本地标签 16023，出标签为 POP 操作，通过指令查看节点 6 的 MPLS 转发表项为：

```
RP/0/0/CPU0:XRV-6#show mpls forwarding
Local    Outgoing    Prefix              Outgoing      Next Hop    Bytes
Label    Label       or ID               Interface                 Switched
------   ---------   -----------------   -----------   ---------   --------
16023    Pop         SR Pfx (idx 23)     Gi0/0/0/0     34.1.1.3    0
```

对于节点 4 而言，它到 Anycast 节点 2 的路径代价为 1，下一跳分别为节点 2 的两个不同接口；到 Anycast 节点 3 的代价为 1，下一跳为节点 3。因此，节点 4 存在 3 条等价路径到达 AnycastIP，MPLS 转发表项为本地标签 16023，出标签为

POP 操作，通过指令查看节点 4 的 MPLS 转发表项为：

```
RP/0/0/CPU0:XRV-4#show mpls forwarding
Local   Outgoing    Prefix           Outgoing      Next Hop    Bytes
Label   Label       or ID            Interface                 Switched
------  ---------   --------------   -----------   ---------   ---------
16023   Pop         SR Pfx (idx 23)  Gi0/0/0/0     34.1.1.3    0
        Pop         SR Pfx (idx 23)  Gi0/0/0/2     24.1.1.2    0
        Pop         SR Pfx (idx 23)  Gi0/0/0/6     24.0.0.2    0
```

对于节点 5 而言，它到节点 2 的路径为节点 5→4→2，到节点 3 的路径为节点 5→4→3 和节点 5→6→3。这意味着节点 5 去往 Anycast 地址存在 3 条等代价路径，但只有两个下一跳。因此，节点 5 将建立起本地标签为 16023，出标签为 16023，且下一跳分别为节点 4 和节点 5 的 MPLS 转发表项，通过指令查看节点 5 的 MPLS 转发表项为：

```
RP/0/0/CPU0:XRV-5#show mpls forwarding
Local   Outgoing    Prefix           Outgoing      Next Hop    Bytes
Label   Label       or ID            Interface                 Switched
------  ---------   --------------   -----------   ---------   ---------
16023   16023       SR Pfx (idx 23)  Gi0/0/0/4     45.1.1.4    0
        16023       SR Pfx (idx 23)  Gi0/0/0/6     56.1.1.6    0
```

下面重点探究一下节点 5 的 Anycast 转发表与节点 2、节点 3 的 AnycastSID 通告之间的关系。如前文所述，节点存在 3 条达到 AnycastIP 的路径，分别为节点 5→4→2、节点 5→4→3 和节点 5→6→3。在计算到达下一跳（节点 4）的出标签时，究竟采用的是节点 5→4→2 还是节点 5→4→3 路径？

之所以探讨这个问题，是因为计算出标签时要使用源发节点的 SID 信息。如果选用节点 5→4→2，则出标签要将节点 2 通告的 SID 索引与节点 4 的 SRGB 相加；如果选用节点 5→4→3，则出标签要将节点 3 通告的 SID 索引与节点 4 的 SRGB 相加。当前的优选规则为"选择通告路由器 ID 大的 SID 索引"，这就意味着节点 4 将使用节点 3 通告的 SID 索引。下面通过实验验证。

最直接的验证方法是在节点 2 和节点 3 上为该 AnycastIP 配置不同的 SID，例如分别将 SID index 设为 23 和 2000。此时所有节点会报告错误信息 "Prefix 23.1.1.23/32 alg 0 assigned SID index 23 by 2.2.2.2 also has index 2000 from 3.3.3.3-

ignoring SID until conflict resolved"。这说明 OSPF 进程会检测 SID 冲突,并将绑定到同一个 IP 上的 SID 设置为失效。

如果先让节点 2 通告 SID 绑定,节点 3 不通告,按照前述规则,节点 5 使用节点 5→4→3 和节点 5→6→3,即使用节点 3 通告 SID 计算 MPLS 转发表项。但节点 3 此时未通告 SID,因此无法建立标签交换表项,出标签被设定为"Unlabelled",即将 MPLS 分组还原为 IP 分组后转发。当携带 16023 的多层标签分组到达节点 5 时,节点 5 将所有标签弹出还原为 IP 报文,并按照 ECMP 方式转发给节点 4 和节点 6。在节点 4 的 3 条等代价路径中,因为选定源自节点 3 的路径为最优路径,所以设置对应的出标签为"Unlabelled"。此时,节点 5 和节点 4 的 MPLS 转发表项如下:

```
RP/0/0/CPU0:XRV-5#show mpls forwarding
Local    Outgoing    Prefix            Outgoing      Next Hop    Bytes
Label    Label       or ID             Interface                 Switched
------   ----------  ----------------  -----------   ---------   --------
16023    Unlabelled  SR Pfx (idx 23)   Gi0/0/0/4     45.1.1.4    0
         Unlabelled  SR Pfx (idx 23)   Gi0/0/0/6     56.1.1.6    0

RP/0/0/CPU0:XRV-4#show mpls forwarding
Local    Outgoing    Prefix            Outgoing      Next Hop    Bytes
Label    Label       or ID             Interface                 Switched
------   ----------  ----------------  -----------   ---------   --------
16023    Unlabelled  SR Pfx (idx 23)   Gi0/0/0/0     34.1.1.3    0
         Pop         SR Pfx (idx 23)   Gi0/0/0/2     24.1.1.2    0
         Pop         SR Pfx (idx 23)   Gi0/0/0/6     24.0.0.2    0
```

如果先让节点 3 通告 SID 绑定,节点 2 不通告,按照前述规则,使用节点 3 通告 SID 计算 MPLS 转发表项,那么节点 5 正常建立 MPLS 转发表项,节点 4 因为其中两条最优路径优选节点 2,设置对应的出标签为"Unlabelled"。此时,节点 5 和节点 4 的 MPLS 转发表项如下:

```
RP/0/0/CPU0:XRV-5#show mpls forwarding
Local    Outgoing    Prefix            Outgoing      Next Hop    Bytes
Label    Label       or ID             Interface                 Switched
------   ----------  ----------------  -----------   ---------   --------
16023    16023       SR Pfx (idx 23)   Gi0/0/0/4     45.1.1.4    0
```

| 16023 | | SR Pfx (idx 23) | Gi0/0/0/6 | 56.1.1.6 | 0 |

```
RP/0/0/CPU0:XRV-4#show mpls forwarding
Local    Outgoing    Prefix              Outgoing     Next Hop   Bytes
Label    Label       or ID               Interface               Switched
------   ----------  -----------------   ----------   --------   --------
16023    Pop         SR Pfx (idx 23)     Gi0/0/0/0    34.1.1.3   0
         Unlabelled  SR Pfx (idx 23)     Gi0/0/0/2    24.1.1.2   0
         Unlabelled  SR Pfx (idx 23)     Gi0/0/0/6    24.0.0.2   0
```

保持节点 3 通告 SID 绑定，节点 2 不通告，将节点的 2 的 RID（Router ID）改为 23.1.1.23，并重启 OSPF 进程。此时，因为节点 2 的 RID 大于节点 3，那么节点 5 针对节点 5→4→2、节点 5→4→3，使用节点 2 通告 SID 计算下一跳（节点 4）的出标签。而节点 2 未通告 SID，因此，下一跳（节点 4）的出标签为 "Unlabelled"。此时，节点 5 和节点 4 的 MPLS 转发表项如下：

```
RP/0/0/CPU0:XRV-5#show mpls forwarding
Local    Outgoing    Prefix              Outgoing     Next Hop   Bytes
Label    Label       or ID               Interface               Switched
------   ----------  -----------------   ----------   --------   --------
16023    Unlabelled  SR Pfx (idx 23)     Gi0/0/0/4    45.1.1.4   0
         16023       SR Pfx (idx 23)     Gi0/0/0/6    56.1.1.6   0

RP/0/0/CPU0:XRV-4#show mpls forwarding
Local    Outgoing    Prefix              Outgoing     Next Hop   Bytes
Label    Label       or ID               Interface               Switched
------   ----------  -----------------   ----------   --------   --------
16023    Pop         SR Pfx (idx 23)     Gi0/0/0/0    34.1.1.3   0
         Unlabelled  SR Pfx (idx 23)     Gi0/0/0/2    24.1.1.2   0
         Unlabelled  SR Pfx (idx 23)     Gi0/0/0/6    24.0.0.2   0
```

5.2.4 数据转发验证

与静态 MPLS 或者 LDP-MPLS 一样，SR-MPLS 仍然可通过 MPLS ping 和 MPLS traceroute 进行路径有效性验证。思科当前版本中提供了对应的 SR-MPLS 指令，即 SR-MPLS ping 与 traceroute。

1. 多路径转发

SR-MPLS 与传统 MPLS 数据转发的关键不同点之一就是支持 MPLS 的 ECMP，以实现端点之间的多条路径资源的充分利用。

当端点之间存在一条路径时，源节点依次发送 TTL 增加的 MPLS Echo 报文，开启 MPLS OAM 服务的节点向源节点回应 TTL 超时报文。由于各节点的下一跳唯一，源节点只需要不断增加发送报文的 TTL 值，并从响应报文中提取回应节点地址即可得到整条路径信息。然而，当节点采用 ECMP 进行数据转发时，各节点去往目的地的下一跳不唯一，它会将流量按照特定哈希规则负载均衡到不同的下一跳节点上。源节点不能仅通过增加 TTL 值实现路径的发现，它必须携带更加详尽的信息，以指导中间节点将探测数据沿着某个路径方向发送。中间节点的 TTL 超时报文也必须携带更多的辅助信息来协助源节点完成上述操作。

Downstream Mapping TLV（DSMAP）用于携带当前节点的多径转发表信息，以辅助源节点进行路径探测，这一个字段被后来的 Detail Downstream Mapping TLV（DDMAP）代替。两者在格式上不同，但承载的信息相差不大，下文围绕 DSMAP 讲解 SR-MPLS 多路径发现过程。

从探测原理来讲，源发节点在发出 TTL=t 的多路径探测包时，携带 M 个目标地址。TTL 超时节点按照其哈希规则和下一跳节点数 N，将这 M 个地址哈希为 N 组，并将每一组地址与转发表项的下一跳、出接口、出标签等信息绑定在一起，即建立第 i 组目标 $IPS_i=\{IP_{i0}, IP_{i1}, \cdots, IP_{ik}\}$ 与 $\{Nexthop_i, Oif_i, Olable_i\}$（$1<i\leqslant N$）的映射关系，此信息被称为 DSMAP 信息。TTL 超时节点通过 MPLS Echo Reply 报文将 DSMAP 信息反馈至源节点。源节点基于这些信息就可以得到特定节点向下一跳转发 MPLS 分组的规则。这样，在进一步的路径探测时，源节点将 TTL 值设置为 $t+1$，并调整目标 IP 地址，将目标 IP 地址设定为第 i 组中的某个地址 IP_{ij}，进而确保新的探测报文到达 TTL=n 的超时节点时，它将该探测分组按照预期转发规则，使用 $Olable_i$ 从 Oif_i 分流到预期的下一跳 $Nexthop_i$。若该探测分组在 $Nexthop_i$ 节点出现 TTL 超时，将生成新的 DSMAP 信息反馈给源节点。源节点通过不停迭代，从而实现多个分流路径的探测发现。

第 5 章　OSPF-SR 原理与实战

下面实验节点 1 探测去往节点 5 的多路径情况。将链路 l_2 和 l_6 的 OSPF 链路代价设置为 2，其他链路代价均为 1。

首先查看节点 1 关于 5.5.5.5/32 的 MPLS 转发表信息，如下：

```
RP/0/0/CPU0:XRV-1#show mpls forwarding prefix 5.5.5.5/32
Local   Outgoing   Prefix            Outgoing      Next Hop    Bytes
Label   Label      or ID             Interface                 Switched
------  ---------  ----------------  ------------  ----------  --------
16005   16005      SR Pfx (idx 5)    Gi0/0/0/2     12.1.1.2    0
        16005      SR Pfx (idx 5)    Gi0/0/0/3     13.1.1.3    520
```

从节点 1 的 MPLS 转发表中可以看到，它拥有两个去往目的地 5.5.5.5/32 的下一跳，分别为节点 2 和节点 3。通过 SR-MPLS traceroute 指令进行路径探测，得到信息如下：

```
RP/0/0/CPU0:XRV-1#traceroute sr-mpls multipath 5.5.5.5/32

LLL!
Path 0 found,
 output interface GigabitEthernet0/0/0/2 nexthop 12.1.1.2
 source 12.1.1.1 destination 127.0.0.3
!
Path 1 found,
 output interface GigabitEthernet0/0/0/2 nexthop 12.1.1.2
 source 12.1.1.1 destination 127.0.0.0
LL!
Path 2 found,
 output interface GigabitEthernet0/0/0/2 nexthop 12.1.1.2
 source 12.1.1.1 destination 127.0.0.9
!
Path 3 found,
 output interface GigabitEthernet0/0/0/2 nexthop 12.1.1.2
 source 12.1.1.1 destination 127.0.0.1
LL!
Path 4 found,
 output interface GigabitEthernet0/0/0/3 nexthop 13.1.1.3
```

```
source 13.1.1.1 destination 127.0.0.0

Paths (found/broken/unexplored) (5/0/0)
Echo Request (sent/fail) (12/0)
Echo Reply (received/timeout) (12/0)
Total Time Elapsed 469 ms
```

从上述结果看出,节点 1 共探测发现了 5 条去往目的节点 5 的路径,其中有 4 条路径经过了下一跳节点 2:Path0、Path1、Path2、Path3。在探测这 4 条路径时,节点 1 使用了连接节点 2 的接口 IP 地址 12.1.1.1 作为源地址,但是目的地址不同,分别为 127.0.0.3、127.0.0.0、127.0.0.9 和 127.0.0.1。在上述过程中,在节点 1 的 Gi0/0/0/2 接口上抓包,得到 SR MPLS traceroute 报文交互过程如图 5.12 所示。

No.	Time	Source	Destination	MPLS-TTL	IP-TTL
132	522.587244	12.1.1.1	127.0.0.0	1	1
133	522.593270	12.1.1.2	12.1.1.1		255
134	522.598769	12.1.1.1	127.0.0.0	2	1
135	522.668582	24.1.1.4	12.1.1.1		254
136	522.676916	12.1.1.1	127.0.0.3	3	1
137	522.710335	34.1.1.6	12.1.1.1		253
138	522.718676	12.1.1.1	127.0.0.3	4	1
139	522.756104	56.1.1.5	12.1.1.1		253
140	522.780042	12.1.1.1	127.0.0.0	3	1
141	522.806629	45.1.1.5	12.1.1.1		253
142	522.816240	12.1.1.1	127.0.0.1		1
143	522.831080	24.0.0.4	12.1.1.1		254
144	522.837669	12.1.1.1	127.0.0.9	3	1
145	522.862067	34.1.1.6	12.1.1.1		253
146	522.896049	12.1.1.1	127.0.0.9	4	1
147	522.913471	56.1.1.5	12.1.1.1		253
148	522.931525	12.1.1.1	127.0.0.1	3	1
149	522.951008	45.1.1.5	12.1.1.1		253

图 5.12 SR MPLS traceroute 报文交互过程

下面结合路径探测过程分析节点 1 选用不同 IP 地址进行路径探索的原因。

第一步,节点 1 发送 TTL=1 的 MPLS Echo Request 报文(序号 132),携带多径信息(IP=127.0.0.0,Mask=0xffffffff),节点 2 将本地 MPLS 分组转发映射规则通过 DSMAP 反馈给节点 1(序号 133)。

拓展前缀 TLV 示例如图 5.13 所示,为节点 2 反馈给节点 1 的 MPLS Echo Reply 报文中 DSMAP 信息。该报文中携带 2 个 DSMAP TLV,说明节点 2 拥有两个等价下一跳,从"Downstream IP Address"获知它们是 24.1.1.4 和 24.0.0.4,

从"Downstream Label Element"域得知出标签为 16005。"Multipath Information"则表明节点 2 在两个下一跳之间的 MPLS 分组转发规则，其携带"IP Adress"（初始 IP）和"Mask"两个字段。前者可理解为"初始 IP 地址"，由探测节点 1 发出 MPLS Echo Request 时设定，后者可理解为"地址增量"，指明当前节点把自给定初始 IP 开始的连续 32 个地址中的哪些绑定相应下一跳上。Mask 长度为 4byte，每一个比特位对应初始 IP 地址。例如，节点 2 反馈的第一条多径信息 Mask 为 0xbd953d04，二进制数为"1011 1101 1001 0101 0011 1101 0000 0100"，按照从左到右顺序，置 1 的比特位数分别为第 0、2、3、4、5、7、8、11、13、15、18、19、20、21、23、29 位。IP 地址组内的 IP 地址计算方法为：初始 IP 地址+比特位数，即 IPS1={127.0.0.0、127.0.0.2、127.0.0.3、127.0.0.4、127.0.0.5、127.0.0.7、127.0.0.8、127.0.0.11、127.0.0.13、127.0.0.15、127.0.0.18、127.0.0.19、127.0.0.20、127.0.0.21、127.0.0.23、127.0.0.29}。通过上述分析可知，目的 IP 地址属于 IPS1 的探测分组将被节点 2 转发给 24.1.1.4。

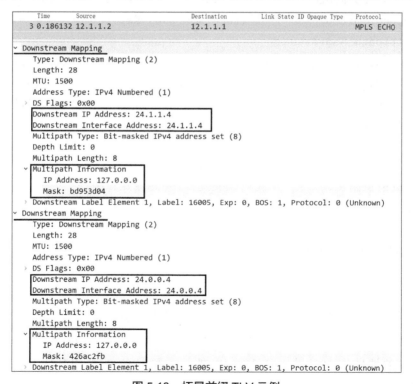

图 5.13 拓展前缀 TLV 示例

采用相同的方法分析节点 2 的第二条路径，Mask=0x426ac2fb=0100 0010 0110 1010 1100 0010 1111 1011，所以，IPS2={127.0.0.1、127.0.0.6、127.0.0.9、127.0.0.10、127.0.0.12、127.0.0.14、127.0.0.16、127.0.0.17、127.0.0.22、127.0.0.24、127.0.0.25、127.0.0.26、127.0.0.27、127.0.0.28、127.0.0.30、127.0.0.31}。目的 IP 地址属于 IPS2 的探测分组将被节点 2 转发给 24.0.0.4。

也可以使用 ping 指令查询路径上某个节点的 DSMAP 信息，例如，节点 1 通过设置 TTL=1 并携带 dsmap 参数，可以让节点 2 反馈其 DSMAP 信息，指令与结果如下：

```
RP/0/0/CPU0:XRV-1#ping sr-mpls 5.5.5.5/32 dsmap ttl 1 output nexthop
12.1.1.2 repeat 1

L Echo Reply received from 12.1.1.2
  DSMAP 0, DS Router Addr 24.1.1.4, DS Intf Addr 24.1.1.4
    Depth Limit 0, MRU 1500 [Labels: 16005 Exp: 0]
    Multipath Addresses:
     127.0.0.0       127.0.0.2       127.0.0.3       127.0.0.4
     127.0.0.5       127.0.0.7       127.0.0.8       127.0.0.11
     127.0.0.13      127.0.0.15      127.0.0.18      127.0.0.19
     127.0.0.20      127.0.0.21      127.0.0.23      127.0.0.29

  DSMAP 1, DS Router Addr 24.0.0.4, DS Intf Addr 24.0.0.4
    Depth Limit 0, MRU 1500 [Labels: 16005 Exp: 0]
    Multipath Addresses:
     127.0.0.1       127.0.0.6       127.0.0.9       127.0.0.10
     127.0.0.12      127.0.0.14      127.0.0.16      127.0.0.17
     127.0.0.22      127.0.0.24      127.0.0.25      127.0.0.26
     127.0.0.27      127.0.0.28      127.0.0.30      127.0.0.31
```

依据节点 2 反馈的下游映射信息，节点 1 即可获知，采用 IPS1 内的地址作为探测分组目的 IP 地址，则节点 2 会将其转发给下一跳 24.1.1.4；采用 IPS2 内的地址作为探测分组目的 IP 地址，则节点 2 会将其转发给下一跳 24.0.0.4。

第二步，节点 1 发送 TTL=2 的 MPLS Echo Request 报文（序号 134），携带多径信息（IP=127.0.0.0，Mask=0xbd953d04），节点 4 将本地 MPLS 分组转发映射规则通过 DSMAP 反馈给节点 1（序号 135）。

TTL=2 的探测报文目的 IP 地址为 127.0.0.0，按照上面分析，节点 2 将该报文转发给 24.1.1.4，并由节点 4 回应其本地 DSMAP，其主要字段信息见表 5.1。

表 5.1 主要字段信息

字段名称	Downstream Mapping1	Downstream Mapping2
Downstream IP Address	34.1.1.6	45.1.1.5
多径信息 --IP Address --Mask	127.0.0.0 0x14152400	127.0.0.0 0xa9801904
Mask 二进制	0001 0100 0001 0101 0010 0100 0000 0000	1010 1001 1000 0000 0001 1001 0000 0100

按照前述规则，节点 4 将 IPS3={127.0.0.3、127.0.0.5、127.0.0.11、127.0.0.13、127.0.0.15、127.0.0.18、127.0.0.21}的分组转发给下一跳 34.1.1.6，将 IPS4={127.0.0.0、127.0.0.2、127.0.0.4、127.0.0.7、127.0.0.8、127.0.0.19、127.0.0.20、127.0.0.23、127.0.0.29}的分组转发给下一跳 45.1.1.5。

对比节点 2 和节点 4 的地址映射信息，两者都拥有两个下一跳，但是节点 4 映射到下一跳的地址数量明显少于节点 2。这是因为在探测过程中，节点在返回多径地址掩码时采用迭代方式，即要将前一个节点的 Mask 与本地 Mask 进行"逻辑与"运算后作为当前节点针对该目的地的分流 Mask。可以直接查询节点 4 关于 5.5.5.5/32 的多径 Mask，如下：

```
RP/0/0/CPU0:XRV-1#ping sr-mpls 5.5.5.5/32 dsmap ttl 2 output nexthop
12.1.1.2 repeat 1 destination 127.0.0.0

L Echo Reply received from 24.1.1.4
  DSMAP 0, DS Router Addr 34.1.1.6, DS Intf Addr 34.1.1.6
    Depth Limit 0, MRU 1500 [Labels: 16005 Exp: 0]
   Multipath Addresses:
     127.0.0.3       127.0.0.5       127.0.0.9       127.0.0.11
     127.0.0.12      127.0.0.13      127.0.0.15      127.0.0.16
```

```
    127.0.0.18      127.0.0.21       127.0.0.22       127.0.0.27
    127.0.0.28      127.0.0.30
DSMAP 1, DS Router Addr 45.1.1.5, DS Intf Addr 45.1.1.5
  Depth Limit 0, MRU 1500 [Labels: explicit-null Exp: 0]
  Multipath Addresses:
    127.0.0.0       127.0.0.1        127.0.0.2        127.0.0.4
    127.0.0.6       127.0.0.7        127.0.0.8        127.0.0.10
    127.0.0.14      127.0.0.17       127.0.0.19       127.0.0.20
    127.0.0.23      127.0.0.24       127.0.0.25       127.0.0.26
    127.0.0.29      127.0.0.31
```

由上述两个地址组可以反推出相应的 Mask 信息，即多径 Mask1=0x145da61a 绑定到下一跳 34.1.1.6；多径 Mask2=0xeba259e5 绑定到下一跳 45.1.1.5。由于节点 1 向节点 4 发送多径请求信息时，携带 Mask=0xbd953d04，该值与 Mask1 和 Mask2 分别进行"逻辑与"运算后得到 0x14152400 和 0xa9801904，这正是节点 4 反馈给节点 1 的多径掩码信息。

第三步，节点 1 发送 TTL=3 的 MPLS Echo Request 报文（序号 136），携带多径信息（IP=127.0.0.0，Mask=0x14152400），节点 6 将本地 MPLS 分组转发映射规则通过 DSMAP 反馈给节点 1（序号 137）。

为了确保 TTL=3 的探测报文经过节点 2、节点 4 后到达节点 6，按照前面探测的结果，必须将目的 IP 地址设置为 IPS3 中的任意地址，这里选用了 127.0.0.3。

由于节点 6 去往 5.5.5.5/32 只有一条路径，故节点 6 反馈给节点 1 的多径掩码信息与节点 4 一样，仍然为 0x14152400，下一跳 IP 为 56.1.1.5。

第四步，节点 1 发送 TTL=4 的 MPLS Echo Request 报文（序号 138），携带多径信息（IP=127.0.0.0，Mask=0x14152400），目的 IP 保持 127.0.0.3 不变。节点 5 在 56.1.1.5 接口收到该分组后，使用该接口地址进行回应，因为它是探测的目标节点，因此，报文（序号 139）中不再携带 DSMAP 信息。

至此，节点 1 已经完成对 Path0 路径的探测，它依次经过节点 2、链路 l_3、节点 4、节点 6 到达节点 5。

由于节点 4 将 IPS4 地址绑定到下一跳（节点 5）上，因此，节点 1 使用目的 IP 127.0.0.0，TTL=3 的探测分组（序号 140）继续探测另一条路径，节点 5

在 45.1.1.5 接口收到该分组后，使用该接口地址进行回应，因为是目标节点，回应报文（序号 141）不携带 DSMAP 信息。

至此，节点 1 已经完成对 Path1 路径的探测，它依次经过节点 2、链路 l_3、节点 4 到达节点 5。

采用类似的方式，节点 1 探测经过节点 2 的链路 l_4 到达节点 5 的不同路径，得到 Path2 和 Path3（序号 142-149）；探测经过节点 3 到达目的节点的路径，得到 Path4。详细路径探测信息如下：

```
RP/0/0/CPU0:XRV-1#traceroute sr-mpls multipath 5.5.5.5/32 verbose
LLL!
Path 0 found,
 output interface GigabitEthernet0/0/0/2 nexthop 12.1.1.2
 source 12.1.1.1 destination 127.0.0.3
  0 12.1.1.1 12.1.1.2 MRU 1500 [Labels: 16005 Exp: 0] multipaths 0
L 1 12.1.1.2 24.1.1.4 MRU 1500 [Labels: 16005 Exp: 0] ret code 8
multipaths 2
L 2 24.1.1.4 34.1.1.6 MRU 1500 [Labels: 16005 Exp: 0] ret code 8
multipaths 2
L 3 34.1.1.6 56.1.1.5 MRU 1500 [Labels: explicit-null Exp: 0] ret
code 8 multipaths 1
! 4 56.1.1.5, ret code 3 multipaths 0
!
Path 1 found,
 output interface GigabitEthernet0/0/0/2 nexthop 12.1.1.2
 source 12.1.1.1 destination 127.0.0.0
  0 12.1.1.1 12.1.1.2 MRU 1500 [Labels: 16005 Exp: 0] multipaths 0
L 1 12.1.1.2 24.1.1.4 MRU 1500 [Labels: 16005 Exp: 0] ret code 8
multipaths 2
L 2 24.1.1.4 45.1.1.5 MRU 1500 [Labels: explicit-null Exp: 0] ret
code 8 multipaths 2
! 3 45.1.1.5, ret code 3 multipaths 0
LL!
Path 2 found,
 output interface GigabitEthernet0/0/0/2 nexthop 12.1.1.2
```

```
  source 12.1.1.1 destination 127.0.0.9
  0 12.1.1.1 12.1.1.2 MRU 1500 [Labels: 16005 Exp: 0] multipaths 0
L 1 12.1.1.2 24.0.0.4 MRU 1500 [Labels: 16005 Exp: 0] ret code 8
multipaths 2
L 2 24.0.0.4 34.1.1.6 MRU 1500 [Labels: 16005 Exp: 0] ret code 8
multipaths 2
L 3 34.1.1.6 56.1.1.5 MRU 1500 [Labels: explicit-null Exp: 0] ret
code 8 multipaths 1
! 4 56.1.1.5, ret code 3 multipaths 0
!
Path 3 found,
 output interface GigabitEthernet0/0/0/2 nexthop 12.1.1.2
 source 12.1.1.1 destination 127.0.0.1
  0 12.1.1.1 12.1.1.2 MRU 1500 [Labels: 16005 Exp: 0] multipaths 0
L 1 12.1.1.2 24.0.0.4 MRU 1500 [Labels: 16005 Exp: 0] ret code 8
multipaths 2
L 2 24.0.0.4 45.1.1.5 MRU 1500 [Labels: explicit-null Exp: 0] ret
code 8 multipaths 2
! 3 45.1.1.5, ret code 3 multipaths 0
LL!
Path 4 found,
 output interface GigabitEthernet0/0/0/3 nexthop 13.1.1.3
 source 13.1.1.1 destination 127.0.0.0
  0 13.1.1.1 13.1.1.3 MRU 1500 [Labels: 16005 Exp: 0] multipaths 0
L 1 13.1.1.3 34.1.1.6 MRU 1500 [Labels: 16005 Exp: 0] ret code 8
multipaths 1
L 2 34.1.1.6 56.1.1.5 MRU 1500 [Labels: explicit-null Exp: 0] ret
code 8 multipaths 1
! 3 56.1.1.5, ret code 3 multipaths 0
```

2. 任意路径转发

SR-MPLS 与传统 MPLS 数据转发的另一个不同点之一就是支持任意路径编排控制。

实验开始前，配置各节点在进行 NodeSID 通告时采用显式空标签，且各链路代价相同。

节点 1 去往目的节点 6 的默认路径为节点 1→3→6，通过 SR-MPLS traceroute 进行探测，得到结果如下：

```
RP/0/0/CPU0:XRV-1#traceroute sr-mpls 6.6.6.6/32 verbose
Sat Jul 31 21:43:36.534 UTC

  0 13.1.1.1 13.1.1.3 MRU 1500 [Labels: 16006 Exp: 0]
L 1 13.1.1.3 34.1.1.6 MRU 1500 [Labels: explicit-null Exp: 0] 20 ms,
ret code 8
! 2 34.1.1.6 10 ms, ret code 3
```

如果期望节点 1 去往节点 6 的数据按照节点 1→3→4→2→4→5→6 进行转发，可通过 NodeSID 进行路径编排。例如，节点 2 发给节点 1 的 MPLS 分组头部依次压入代表路径上各节点的标签，从栈顶到栈底依次为：0, 16003, 16004, 16002, 16004, 16005, 16006。因为所有节点均采用了显式空标签，节点 2 作为节点 1 的倒数第二跳，故使用标签 0 作为栈顶标签。由于采用了自编排路径，因此，要使用 SR-MPLS 的 Nil-FEC 进行路径测试与验证，探测指令与结果如下：

```
RP/0/0/CPU0:XRV-2#traceroute sr-mpls nil-fec labels 0, 16003, 16004,
16002, 16004, 16005, 16006 output interface g0/0/0/1 nexthop
12.1.1.1 verbose

  0 12.1.1.2 12.1.1.1 MRU 1500 [Labels: explicit-null/16003/16004/
16002/16004/16005/ 16006/explicit-null Exp: 0/0/0/0/0/0/0/0]
L 1 12.1.1.1 13.1.1.3 MRU 1500 [Labels: explicit-null/16004/16002/
16004/16005/ 16006/explicit-null Exp: 0/0/0/0/0/0/0] 30 ms, ret code 8
L 2 13.1.1.3 34.1.1.4 MRU 1500 [Labels: explicit-null/16002/16004/
16005/16006/explicit-null Exp: 0/0/0/0/0/0] 10 ms, ret code 8
L 3 34.1.1.4 24.0.0.2 MRU 1500 [Labels: explicit-null/16004/16005/
16006/explicit-null Exp: 0/0/0/0/0] 10 ms, ret code 8
L 4 24.0.0.2 24.1.1.4 MRU 1500 [Labels: explicit-null/16005/16006/
explicit-null Exp: 0/0/0/0] 10 ms, ret code 8
```

```
L 5 24.1.1.4 45.1.1.5 MRU 1500 [Labels: explicit-null/16006/explicit-
null Exp: 0/0/0] 10 ms, ret code 8
L 6 45.1.1.5 56.1.1.6 MRU 1500 [Labels: explicit-null/explicit-null
Exp: 0/0] 20 ms, ret code 8
! 7 56.1.1.6 10 ms, ret code 3
```

除了使用 NodeSID 进行路径编排实现流量引导，也可以采用 AdjSID 进行流量引导。可以查看每个节点为其直连链路分配的标签信息，假定期望流量依次经过链路 l_2、l_5、l_3、l_4、l_6、l_7，那么可以使用的标签栈信息为{30002, 24000, 24002, 24002, 24003, 24001}，探测指令与结果如下：

```
RP/0/0/CPU0:XRV-2#traceroute sr-mpls nil-fec labels 30002,24000,24002,
24002,24003,24001 output interface g0/0/0/1 nexthop 12.1.1.1 verbose

  0 12.1.1.2 12.1.1.1 MRU 1500 [Labels: 30002/24000/24002/24002/24003/
24001/explicit-null Exp: 0/0/0/0/0/0/0]
L 1 12.1.1.1 13.1.1.3 MRU 1500 [Labels: implicit-null/24000/24002/
24002/24003 /24001/explicit-null Exp: 0/0/0/0/0/0/0] 10 ms, ret code 8
L 2 13.1.1.3 34.1.1.4 MRU 1500 [Labels: implicit-null/24002/24002/
24003/24001/explicit-null Exp: 0/0/0/0/0/0] 10 ms, ret code 8
L 3 34.1.1.4 24.1.1.2 MRU 1500 [Labels: implicit-null/24002/24003/
24001/explicit-null Exp: 0/0/0/0/0] 10 ms, ret code 8
L 4 24.1.1.2 24.0.0.4 MRU 1500 [Labels: implicit-null/24003/24001/
explicit-null Exp: 0/0/0/0] 10 ms, ret code 8
L 5 24.0.0.4 45.1.1.5 MRU 1500 [Labels: implicit-null/24001/explicit-
null Exp: 0/0/0] 10 ms, ret code 8
L 6 45.1.1.5 56.1.1.6 MRU 1500 [Labels: implicit-null/explicit-null
Exp: 0/0] 10 ms, ret code 8
! 7 56.1.1.6 20 ms, ret code 3
```

单纯使用 NodeSID 或者 AdjSID 实现路径编排，往往会使得标签栈过深，而一些硬件设备支持的标签栈深度是有限的。因此，在实际应用中应当将两者进行组合使用，以更为精简的标签栈实现同样的流量引导效果。例如，为了实现上面实验中纯粹使用 AdjSID 实现引流的效果，可以将 NodeSID 和 AdjSID 混编为{30002,16004,24002,24002,16005,16006}，探测指令与结果如下：

第5章 OSPF-SR 原理与实战

```
RP/0/0/CPU0:XRV-2#traceroute sr-mpls nil-fec labels 30002,16004,24002,
24002,16005,16006 output interface g0/0/0/1 nexthop 12.1.1.1 verbose

 0 12.1.1.2 12.1.1.1 MRU 1500 [Labels: 30002/16004/24002/24002/16005/
16006/explicit-null Exp: 0/0/0/0/0/0/0]
L 1 12.1.1.1 13.1.1.3 MRU 1500 [Labels: implicit-null/16004/24002/
24002/16005 /16006/explicit-null Exp: 0/0/0/0/0/0/0] 20 ms, ret code 8
L 2 13.1.1.3 34.1.1.4 MRU 1500 [Labels: explicit-null/24002/24002/
16005/16006/explicit-null Exp: 0/0/0/0/0/0] 10 ms, ret code 8
L 3 34.1.1.4 24.1.1.2 MRU 1500 [Labels: implicit-null/24002/16005/
16006/explicit-null Exp: 0/0/0/0/0] 10 ms, ret code 8
L 4 24.1.1.2 24.0.0.4 MRU 1500 [Labels: implicit-null/16005/16006/
explicit-null Exp: 0/0/0/0] 0 ms, ret code 8
L 5 24.0.0.4 45.1.1.5 MRU 1500 [Labels: explicit-null/16006/explicit-
null Exp: 0/0/0] 0 ms, ret code 8
L 6 45.1.1.5 56.1.1.6 MRU 1500 [Labels: explicit-null/explicit-null
Exp: 0/0] 10 ms, ret code 8
! 7 56.1.1.6 20 ms, ret code 3
```

5.3 域间 SR 实验

区域间相互通告前缀信息，相应 PrefixSID 也伴随着在区域之间洪泛。域间 SR 实验主要围绕 PrefixSID 的跨区域传播展开。

5.3.1 实验环境

多区域 SR 实验拓扑如图 5.14 所示，在图 5.9 的基础上增加了 3 个节点（即节点 7、节点 8、节点 9），所有路由器运行了 OSPF 协议，节点 1~6 位于区域 0 中，节点 7、节点 8 位于区域 78，节点 9 位于区域 9。节点 1~节点 6 的 SRGB 为[16000, 23999]，节点 7、节点 8、节点 9 的 SRGB 分别为[70000, 71000]、[80000, 81000]、[90000, 91000]。所有 OSPF 链路代价均为 1。在节点 N 设置环回接口，并配置 IP

地址 N.N.N.N/32。各节点通过 OSPF 协议交互网络拓扑信息，建立到所有接口网段和主机地址的路由。

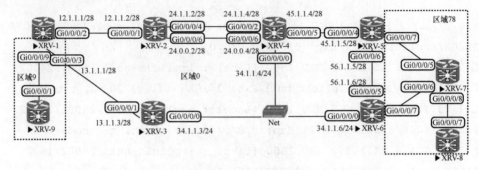

图 5.14 多区域 SR 实验拓扑

5.3.2 协议交互

1. 消息通告内容的变化

启动 SR 功能的区域边界路由器（Area Border Router，ABR）将区域内拓展前缀 LSA 转换为区域间拓展前缀 LSA 进行洪泛。在跨区域传播时保持前缀和 SID 不变，仅在某些内容上修改以满足消息传播和转发表项计算的约束。以节点 8 通告的拓展前缀 LSA 为例，节点 6 收到和中继的拓展前缀 LSA 消息如图 5.15 所示，其中右侧为从区域 78 收到的消息，左侧为转发到区域 0 的 LSA 消息。

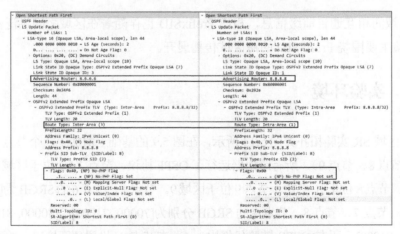

图 5.15 节点 6 收到和中继的拓展前缀 LSA 消息

通过对比可以看到,在消息内容上存在以下几点变化:

1)通告路由器 ID 字段由 8.8.8.8 修改为 6.6.6.6,用于告知区域 0 内的路由器,该 PrefixSID 为节点 6 始发,区域 0 内节点将基于该条件进行转发表的计算;

2)拓展前缀 TLV 中的路由类型由 1 改为 3,表示区域间路由,这与域间路由扩散时将 LSA1 转为 LSA3 的操作类似;

3)NP 与 E 标志位发生改变,以辅助其他节点正确地计算转发表。

2. NP 和 E 标志位的变化

1)中继 NP=0 且 E=0 消息时将 NP 比特置 1

E=0 意味着源发节点不支持携带显式空标签的 MPLS 分组处理,并通过 NP=0 要求倒数第二跳节点执行标签弹出操作。收到携带这组标志位的节点,依据当前路由判定自己为倒数第二跳节点后,安装执行 POP 操作的 MPLS 转发表项。节点 7 关于 8.8.8.8/32 的 MPLS 转发表如下:

```
RP/0/0/CPU0:XRV-7#show mpls forwarding prefix 8.8.8.8/32
Local   Outgoing    Prefix            Outgoing       Next Hop    Bytes
Label   Label       or ID             Interface                  Switched
------  ----------  ----------------  -------------  ----------  ----------
70008   Pop         SR Pfx (idx 8)    Gi0/0/0/8      78.1.1.8    0
```

由于 ABR 在中继 PrefixSID 时将通告路由器 ID 修改为自身 ID,这使得当前域的节点在判定倒数第二跳时产生了假象。例如,节点 4 通告 8.8.8.8 的 PrefixSID 时将通告路由器设定为自己,使得节点 5 认为节点 4 直连 8.8.8.8,于是认定自己为去往 8.8.8.8 的倒数第二跳,并按照 NP=0 且 E=0 标志位安装标签弹出表项,这样使得段标签在没有进入目的区域时就被弹出了,无法达到 SID 引导流量的目的。因此,ABR 在中继 PrefixSID 时将 NP 比特置 1,确保倒数第二跳节点不执行标签弹出操作,而执行标签交换操作,从而保留了 SID 标签信息,这样后续转发能够继续按照 SID 引导流量。例如,节点 4 虽然识别自己为倒数第二跳节点,但是仍然安装 SWAP 操作的 MPLS 转发表项,确保去往 8.8.8.8 的分组能够继续携带标签到达真正的倒数第二跳(节点 7)后再执行标签弹出操作。

```
RP/0/0/CPU0:XRV-4#show mpls forwarding prefix 8.8.8.8/32 detail
Local   Outgoing    Prefix            Outgoing       Next Hop    Bytes
Label   Label       or ID             Interface                  Switched
------  ----------  ----------------  -------------  ----------  ----------
```

```
16008      16008           SR Pfx (idx 8)    Gi0/0/0/0       34.1.1.6    0
    Label Stack (Top -> Bottom): { 16008 }
    NHID: 0x0, Encap-ID: N/A, Path idx: 0, Backup path idx: 0,
Weight: 0
    MAC/Encaps: 14/18, MTU: 1500
    Outgoing Interface: GigabitEthernet0/0/0/0 (ifhandle 0x00000040)
    Packets Switched: 0

    16008           SR Pfx (idx 8)    Gi0/0/0/5       45.1.1.5    0
    Label Stack (Top -> Bottom): { 16008 }
    NHID: 0x0, Encap-ID: N/A, Path idx: 1, Backup path idx: 0,
Weight: 0
    MAC/Encaps: 14/18, MTU: 1500
    Outgoing Interface: GigabitEthernet0/0/0/5 (ifhandle 0x000000e0)
    Packets Switched: 0
```

2）中继 NP=1 且 E=1 消息时将 E 比特置 0

E=1 意味着源发节点支持携带显式空标签的 MPLS 分组处理，并通过 NP=1 要求倒数第二跳节点执行标签 SWAP 操作。收到携带这组标志位的节点，依据当前路由判定自己为倒数第二跳节点后，安装执行 SWAP 操作的 MPLS 转发表项，将最外层标签交换为标签 0。例如，配置节点 8，在通告 PrefixSID 时采用显式空标签，指令为：

```
router ospf 1 area 78 interface loopback0 prefix-sid index 8 explicit-
null
```

此时节点 8 和节点 7 的 MPLS 转发表如下：

```
RP/0/0/CPU0:XRV-7#show mpls forwarding prefix 8.8.8.8/32
Local   Outgoing      Prefix            Outgoing      Next Hop     Bytes
Label   Label         or ID             Interface                  Switched
------  ------------  ----------------  ------------  -----------  --------
70008   Exp-Null-v4   SR Pfx (idx 8)    Gi0/0/0/8     78.1.1.8     0
```

同样，由于 ABR 在中继 PrefixSID 时修改通告路由器 ID，会产生倒数第二跳假象。例如，当节点 4 认定自己为去往 8.8.8.8 的倒数第二跳，并按照 NP=1 且 E=1 标志位安装标签交换表项，将栈顶标签交换为标签 0，如果 ABR 节点没有安装标签 0 的表项，那么将导致 SID 转发失败。因此，ABR 在中继 PrefixSID

时将 E 比特置 0，告知倒数第二跳节点自己不支持显式空标签，倒数第二跳节点不能将栈顶标签 SWAP 为 0，而是执行正常标签 SWAP 操作，从而保留了 SID 标签信息。节点 6 收到（右）和中继（左）的拓展前缀 LSA 消息如图 5.16 所示。

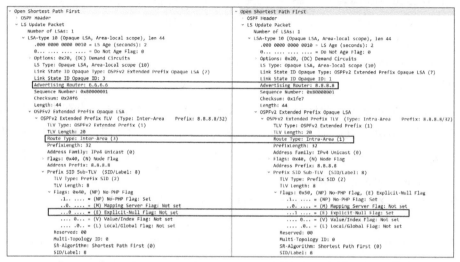

图 5.16　节点 6 收到（右）和中继（左）的拓展前缀 LSA 消息

例如，节点 4 针对 8.8.8.8 安装标签信息如下：

```
RP/0/0/CPU0:XRV-4#show mpls forwarding prefix 8.8.8.8/32
Local    Outgoing    Prefix            Outgoing      Next Hop      Bytes
Label    Label       or ID             Interface                   Switched
------   --------    ---------------   -----------   -----------   --------
16008    16008       SR Pfx (idx 8)    Gi0/0/0/0     34.1.1.6      0
         16008       SR Pfx (idx 8)    Gi0/0/0/5     45.1.1.5      0
```

通过修改 E、NP 标志，控制沿途中的所有节点不要提前执行标签弹出或者不要交换为标签 0，而是执行正常标签交换，从而在跨域转发时一致保留 SID 信息，进而实现基于 SID 的流量引导目的。

3．A 标志位作用与变化

存在 ABR 节点将自己直连前缀从一个区域传播到另一个区域的情况，在这种情况下，ABR 节点既是中继节点又是前缀源发节点，因此，ABR 修改通告路由器 ID 不会对其他节点产生倒数第二跳假象，ABR 在中继消息时不需要修改 NP、E 字段。为了与其他 PrefixSID 区分，通告消息中要携带 Attach 标志，以标识该前缀直接挂载在 ABR 节点上。

在节点 5 上配置一个 55.55.55.55/32 的环回接口,将其置于区域 78,分配 PrefixSID 55,此时节点 4 的 MPLS 转发表项如下:

```
RP/0/0/CPU0:XRV-4#show mpls forwarding prefix 55.55.55.55/32
Local    Outgoing    Prefix             Outgoing      Next Hop    Bytes
Label    Label       or ID              Interface                 Switched
------   ----------  ----------------   ------------  ----------  --------
16055    Pop         SR Pfx (idx 55)    Gi0/0/0/5     45.1.1.5    0
```

然而,A 标志不能跨区域传播。收到携带 A 标志的 ABR 节点在中转时仍需要修改通告路由器 ID,此时 A 标志必须置 0 以反映真实情况,NP 和 E 标志位也需要按照前述规则进行对应修改,以辅助跨域建立正确 MPLS 转发表项。此时节点 9 的 MPLS 转发表项如下:

```
RP/0/0/CPU0:XRV-9#show mpls forwarding prefix 55.55.55.55/32
Local    Outgoing    Prefix             Outgoing      Next Hop    Bytes
Label    Label       or ID              Interface                 Switched
------   ----------  ----------------   ------------  ----------  --------
90055    16055       SR Pfx (idx 55)    Gi0/0/0/1     19.1.1.1    0
```

注:A 标志在当前版本上并未被支持,可能会在更高级版本上予以实现。

4. 跨域标签转发路径的建立

由于 ABR 在通告区域 PrefixSID 时修改了通告路由器 ID 字段,区域内节点计算到达其他区域 Prefix 的 MPLS 转发表项时,仅依赖当前区域的 ABR 信息,而不依赖其他区域信息。即使携带 SRGB 的路由器 ID 信息只在区域内部传播,也不会影响整体转发路径的建立。

同时,因为 ABR 节点跨越在区域的边界,拥有相关区域内完整的信息建立 MPLS 转发表项。于是,区域内节点依据 ABR 建立了去往其他区域的转发信息,自动实现了不同区域之间标签的映射绑定。节点 9 位于区域 9 中,虽然没有接收到区域 0 和区域 78 内 PrefixSID 信息,但可以依托 ABR 节点 1 建立起通往所有这些节点的转发路径,其 MPLS 转发表信息如下:

```
RP/0/0/CPU0:XRV-9#show mpls forwarding
Local    Outgoing    Prefix             Outgoing      Next Hop    Bytes
Label    Label       or ID              Interface                 Switched
```

```
------  --------------  ---------------  ----------   ----------  ------
90001   Exp-Null-v4     SR Pfx (idx 1)   Gi0/0/0/1    19.1.1.1    0
90002   16002           SR Pfx (idx 2)   Gi0/0/0/1    19.1.1.1    0
90003   16003           SR Pfx (idx 3)   Gi0/0/0/1    19.1.1.1    0
90004   16004           SR Pfx (idx 4)   Gi0/0/0/1    19.1.1.1    0
90005   16005           SR Pfx (idx 5)   Gi0/0/0/1    19.1.1.1    0
90006   16006           SR Pfx (idx 6)   Gi0/0/0/1    19.1.1.1    0
90007   16007           SR Pfx (idx 7)   Gi0/0/0/1    19.1.1.1    0
90008   16008           SR Pfx (idx 8)   Gi0/0/0/1    19.1.1.1    0
```

5.3.3 数据转发验证

通过 ABR 的中继，网络中节点都能获知 OSPF 域内任意节点的 PrefixSID index，然而，由于携带 SRGB 的 RI LSA 不能跨区域传送，所以，一个区域内的节点无法获知其他区域节点的 PrefixSID，也就无法进行有效的路径控制和流量引导。例如，上表中，节点 9 虽然建立了去往所有节点的 MPLS 转发表项，却无法得知每一个节点的 PrefixSID 信息，这与 SR 设计理念中"PrefixSID 是局部含义，而 SID Index 是全局含义"是一致的。

在节点 9 可以通过路径追踪方式探测去往节点 8 的 MPLS 路径信息，如下：

```
RP/0/0/CPU0:XRV-9#traceroute mpls ipv4 8.8.8.8/32 verbose
  0 19.1.1.9 19.1.1.1 MRU 1500 [Labels: 16008 Exp: 0]
L 1 19.1.1.1 13.1.1.3 MRU 1500 [Labels: 16008 Exp: 0] 0 ms, ret code 8
L 2 13.1.1.3 34.1.1.6 MRU 1500 [Labels: 16008 Exp: 0] 10 ms, ret code 8
L 3 34.1.1.6 67.1.1.7 MRU 1500 [Labels: 70008 Exp: 0] 10 ms, ret code 8
L 4 67.1.1.7 78.1.1.8 MRU 1500 [Labels: explicit-null Exp: 0] 10 ms,
ret code 8
! 5 78.1.1.8 10 ms, ret code 3
```

可以看到去往节点 8 的路径依次经过区域 0 的节点 1、节点 3、节点 6 和区域 78 的节点 7 和节点 8，跨区域路径正常建立成功。

受限于区域内节点只有区域拓扑视图，单纯依赖区域节点获取的信息无法实现跨区域流量调度。因此，在跨域场景中，需要引入集中控制器，控制器通过接入不同的区域来获得每个区域的拓扑、节点层面的 PrefixSID 信息，甚至链路层面 AdjSID 信息。通过全局视角进行流量感知和基于 SID 的流量牵引，进而实现跨域路径编排与调度。

5.4 外部 SR 实验

通过路由重发布的方式，不同协议之间可以实现路由的双向互通。同样，通过路由重发布，也可以在不同协议之间交互 PrefixSID 信息。

通过重发布操作实现了两个不同的协议进程之间路由和 PrefixSID 的互通。为了确保数据面操作的一致性，要求重发布的两个协议进程必须采用相同的 SRGB。如果这个条件不满足，那么只有路由被重发布，而 PrefixSID 则不会。

5.4.1 实验环境

多区域 SR 实验拓扑如图 5.17 所示，节点 1 和节点 9 之间运行 BGP，节点 9 属于 AS900，其他节点属于 AS100。包括节点 1 在内的其他节点运行 OSPF 协议，节点 1～节点 6 位于区域 0，节点 7、节点 8 位于区域 78。节点 1～节点 6 的 SRGB 为[16000, 23999]，节点 7、节点 8 的 SRGB 分别为[70000, 71000]、[80000, 81000]。所有 OSPF 链路代价均为 1。在节点 N 设置环回接口，配置 IP 地址 N.N.N.N/32，分配 SID 索引 N。AS100 内各节点通过 OSPF 协议交互网络拓扑信息，建立到所有接口网段、主机地址的路由以及 MPLS 转发表项。

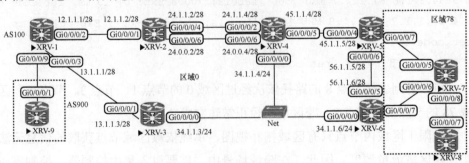

图 5.17 多区域 SR 实验拓扑

5.4.2 协议配置

在节点 9 配置 BGP，并通过路由策略方式为本地环回地址 9.9.9.9/32 设置 PrefixSID 索引 9；然后，启动 BGP LU 地址族，通过 LU 携带 BGP PrefixSID 属性实现 SID 索引通告。节点 9 的 BGP 相关配置如下：

```
route-policy SID($SID)
  set label-index $SID
end-policy
!
route-policy PASS_ALL
  pass
end-policy
!
router bgp 900
 address-family ipv4 unicast
 network 9.9.9.9/32 route-policy SID(9)//本地环回地址设置PrefixSID索引
  allocate-label all
 !
 neighbor 19.1.1.1
  remote-as 100
  address-family ipv4 labeled-unicast//与节点1使能BGP LU地址族
   route-policy PASS_ALL in
   route-policy PASS_ALL out
  !
 !
!
```

必须在 LU 地址族下配置 route-policy 才能在邻居之间使能 BGP LU，否则，BGP LU 将无法启动。

采用类似方法配置节点 1：

```
route-policy PASS_ALL
  pass
```

```
end-policy
!
router bgp 100
 address-family ipv4 unicast
  allocate-label all
 !
 neighbor 19.1.1.9
  remote-as 900
  address-family ipv4 labeled-unicast
   route-policy PASS_ALL in
   route-policy PASS_ALL out
  !
 !
!
segment-routing
 global-block 16000 23999
!
```

配置完上述指令后，在节点 1 查看 BGP LU 信息，如下：

```
RP/0/0/CPU0:XRV-1#show bgp ipv4 labeled-unicast 9.9.9.9/32 detail
Paths: (1 available, best #1)
  Not advertised to any peer
  Path #1: Received by speaker 0
  Flags: 0x4080000009060001, import: 0x20
  Not advertised to any peer
  900
    19.1.1.9 from 19.1.1.9 (9.9.9.9)
      Received Label 3
      Origin IGP, metric 0, localpref 100, valid, external, best, group-best, labeled-unicast
      Received Path ID 0, Local Path ID 0, version 5
      Origin-AS validity: not-found
      Prefix SID Attribute Size: 10
      Label Index: 9
```

可以看到节点 1 通过 BGP LU 从节点 9 收到了关于 9.9.9.9/32 的标签分发消息，分配传统标签 3，同时携带 SID 索引 9。查看节点 1 建立的 MPLS 转发表项信息，如下：

```
RP/0/0/CPU0:XRV-1#show mpls forwarding prefix 9.9.9.9/32
Local   Outgoing   Prefix            Outgoing      Next Hop    Bytes
Label   Label      or ID             Interface                 Switched
------  ---------  ----------------  ------------  ----------  --------
16009   Pop        SR Pfx (idx 9)                  19.1.1.9    0
```

由于 BGP-LU 不会迭代直连路由，此时 MPLS 转发表项的出接口为空，没有建立起正常的 MPLS 转发表项，必须手动配置通往邻居的静态路由。以节点 1 为例，需要手动配置下一跳和出接口的绑定关系。

```
router static
 address-family ipv4 unicast
  19.1.1.9/32 GigabitEthernet0/0/0/9
```

此时，节点 1 建立 MPLS 转发表项，因为 BGP 不通告 SRGB，这里的本地标签基于本地配置的 SRGB 和 BGP 邻居通告的 SID 索引计算得到，正是由于该限制，要求所有 BGP 节点使用相同的 SRGB。

```
RP/0/0/CPU0:XRV-1#show mpls forwarding prefix 9.9.9.9/32
Local   Outgoing   Prefix            Outgoing      Next Hop    Bytes
Label   Label      or ID             Interface                 Switched
------  ---------  ----------------  ------------  ----------  --------
16009   Pop        SR Pfx (idx 9)    Gi0/0/0/9     19.1.1.9    0
```

5.4.3 协议交互

接下来配置节点 1，将 BGP 通告的路由和标签信息发布到 OSPF 中，配置如下：

```
router ospf 1 redistribute bgp 100 tag 900
```

在完成上述配置的同时，在节点 1 的 Gi0/0/0/2 接口和区域 78 的节点 8 接口抓取协议分组，在区域 0 和区域 78 洪泛的拓展前缀信息如图 5.18 所示，左侧为节点 1 发出的拓展前缀信息，右侧为节点 8 收到的拓展前缀信息。从消息中看到，节点 1 作为自治系统边界路由器（Autonomous System Boundary Router，ASBR），将 BGP 通告的 PrefixSID 信息通过 LSA11 承载，并在整个 OSPF 洪泛，节点 1 和节点 8 虽然不在同一 OSPF 区域，但收到的 LSA11 信息是一样的。从 LSA11 消息内容来看，无法识别该拓展前缀来自哪个外部路由协议。

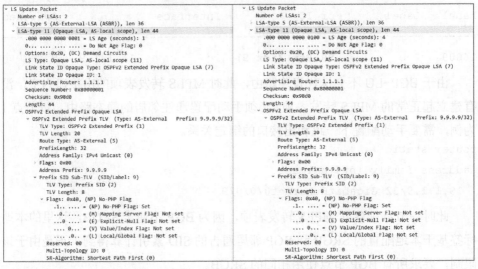

图 5.18　在区域 0 和区域 78 洪泛的拓展前缀信息

对于 ASBR 节点而言，从其他协议学习到的路由为非本地路由，为了避免其他节点产生虚假倒数第二跳，一定要告知与 ASBR 直连的节点不能执行倒数第二跳弹出，因此，在通告中它将 NP 标志位置 1，而且在整个 OSPF 洪泛过程中，该标志都不会发生变化。

所有区域内的 OSPF 节点都将基于 LSA11 中的信息计算得到去往 BGP 路由 9.9.9.9/32 的 MPLS 转发表信息，以节点 8 为例，它识别出节点 1.1.1.1 通告 9.9.9.9/32 路由，而本地到达 1.1.1.1 的下一跳为 78.1.1.7（节点 7）。

```
RP/0/0/CPU0:XRV-8#show ip route
O IA 1.1.1.1/32 [110/5] via 78.1.1.7, 06:01:31, GigabitEthernet0/0/0/7
```

于是出标签为节点 7 的 SRGB+SID=70000+9=70009，而入标签则由本地 SRGB 决定，即 80000+9=80009，于是得到节点 8 的 MPLS 转发表项。

```
RP/0/0/CPU0:XRV-8#show mpls forwarding prefix 9.9.9.9/32
Local   Outgoing   Prefix             Outgoing       Next Hop      Bytes
```

```
Label    Label    or ID              Interface         Switched
------   ------   ---------------    -------------     ---------
80009    70009    SR Pfx (idx 9)     Gi0/0/0/7         78.1.1.7         0
```

5.4.4 数据转发验证

现在已经实现了外部路由在 OSPF 域的传播，各节点也已经建立了去往 BGP 路由的转发表，例如，在节点 8 上通过 MPLS traceroute 可以发现单项路径已经连通。

```
RP/0/0/CPU0:XRV-8#traceroute mpls ipv4 9.9.9.9/32 verbose

  0 78.1.1.8 78.1.1.7 MRU 1500 [Labels: 70009 Exp: 0]
L 1 78.1.1.7 67.1.1.6 MRU 1500 [Labels: 16009 Exp: 0] 10 ms, ret code 8
L 2 67.1.1.6 34.1.1.3 MRU 1500 [Labels: 16009 Exp: 0] 10 ms, ret code 8
L 3 34.1.1.3 13.1.1.1 MRU 1500 [Labels: 16009 Exp: 0] 10 ms, ret code 8
L 4 13.1.1.1 19.1.1.9 MRU 1500 [Labels: implicit-null/implicit-null
Exp: 0/0] 10 ms, ret code 8
. 5 *
. 6 *
```

从路径追踪结果可以看到，探测分组已经到达节点 9，由于节点 1 没有将 OSPF 路由发布到 BGP 中，位于 BGP 域的节点 9 没有得到 AS100 内所有节点的路由信息，也就没有去往节点 8 的路由信息，导致数据面未连通。通过在节点 1 的 BGP 进程中将 OSPF 路由重发布，就可以实现节点 8 和节点 9 之间跨域 SR 互通。

第 6 章

IS-IS SR 原理与实战

6.1 IS-IS SR 原理

6.1.1 IS-IS 链路状态信息与 SID 的关联

与 OSPF 不同，IS-IS 协议通告的链路状态信息 LSP 采用 TLV 格式，易于扩展。添加新的功能时，只需要定义新的 TLV 或子 TLV 类型即可。

IS-IS 路径元素共 5 种，包括实体节点、虚拟节点、前缀、物理链路和逻辑链路。可将这 5 种路径元素概括为两类：一类为前缀标识方式，包括实体节点、虚拟节点以及前缀，它们可统一采用"前缀+掩码"表示；另一类为链路标识方式，包括物理链路和逻辑链路。SID 的分配就是将 MPLS 标签绑定到相应前缀或者链路上的过程。

对于前缀标识，IS-IS 的每个路由前缀信息都对应 LSP 报文里的一个 Extended IP Reachability TLV。IS-IS 通过在指定前缀的 TLV 里添加 PrefixSID Sub-TLV，携带前缀标识信息。

对于链路标识，IS-IS 的每条链路信息都对应 LSP 报文里的一个 Extended IS Reachability TLV。IS-IS 通过在指定链路的 TLV 里添加 AdjSID Sub-TLV 或 LAN-AdjSID Sub-TLV 来包含链路标识信息。

IS-IS 协议里的 Extended IP Reachability TLV 和 Extended IS Reachability TLV 的具体格式可参阅下文中的实验内容。

6.1.2 SID 信息的洪泛方式

IS-IS 协议设计原则规定 2 个 L1 区域之间不能直接通信，必须经过 L2 骨干网。通常区域边界 L1/2 路由器会通过设置 ATT 位，让 L1 路由器建立到自己的默认路由。

L1 LSP 只能在一个 L1 区域内部传播，L2 LSP 只能在 L2 骨干网内传播。IS-IS 默认会把 L1 区域的路由信息，包括 PrefixSID 自动汇入 L2 层路由，通过区域边界 L1/2 路由器传播到骨干网。

L1 区域通常不知道 L2 骨干网的拓扑信息，包括骨干网的 PrefixSID。在某些情况下，如区域存在多个区域边界 L1/2 路由器时，L1/2 路由器也可以配置把 L2 路由信息注入 L1 路由区域，优化 L1 路由选择出口，此时 L1 区域可以知道骨干网的 PrefixSID。

IS-IS 协议也可以通过路由重新发布的方式将外部路由信息和对应的 PrefixSID 注入 L1/L2 路由区域中，其传播方式和普通 PrefixSID 相同。

所有的链路 SID 都只会在对应的区域内洪泛。

IS-IS 的 SID 信息遵从 IS-IS 的消息传播机制，节点按照前述约束条件对收到的 SID（标签映射）消息进行洪泛。这种方法使得域内每个节点都持有相同的 SID 信息，这就意味着每个节点都掌握了相同的域内路径元素信息，为域内路径的任意编排提供了基础支撑。

6.2 域内 SR 实验

本节主要讲解 IS-IS 协议在一个单一区域中，传播 SR 相关信息的过程，包括 SR 能力的通告、各类 SID 信息的洪泛过程等。同时验证在单一区域中，利用 SR 标签进行数据转发的能力。

6.2.1 实验环境

域内 SR 实验拓扑如图 6.1 所示，所有路由器运行了 IS-IS 协议，并处于同一个 L2 区域中，节点 3、节点 4、节点 6 接入 LAN，其他节点之间为点到点链路类型，所有 IS-IS 链路代价均默认为 10。在节点 N 设置环回接口，并配置 IP 地址 $N.N.N.N/32$。各节点通过 IS-IS 协议交互网络拓扑信息，建立到所有接口网段和主机地址的路由。

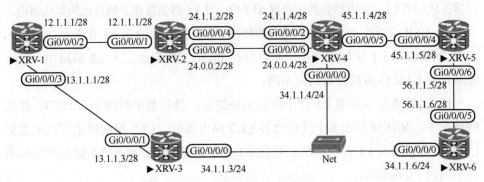

图 6.1 域内 SR 实验拓扑

6.2.2 协议配置

按照实验要求，在每个节点配置 IS-IS 协议，在各互联接口、本地环回接口上启动 IS-IS。

SR 在为本地路由分配标签时，要从本地标签池中申请标签，为了简化，节点 N 的本地标签池为 $N00000\sim N99999$。各节点协议配置信息如下：

```
节点 1:
router isis 1
 is-type level-2-only
 net 10.0000.0000.0001.00
 address-family ipv4 unicast
  metric-style wide
 !
 interface Loopback1
  address-family ipv4 unicast
  !
 !
 interface GigabitEthernet0/0/0/2
  point-to-point
  address-family ipv4 unicast
  !
```

```
节点 2:
router isis 1
 is-type level-2-only
 net 10.0000.0000.0002.00
 address-family ipv4 unicast
  metric-style wide
 !
 interface Loopback1
  address-family ipv4 unicast
  !
 !
 interface GigabitEthernet0/0/0/1
  point-to-point
  address-family ipv4 unicast
  !
```

```
!                                          interface GigabitEthernet0/0/0/4
interface GigabitEthernet0/0/0/3            point-to-point
 point-to-point                             address-family ipv4 unicast
 address-family ipv4 unicast               !
!                                          !
!                                          interface GigabitEthernet0/0/0/6
mpls oam                                    point-to-point
!                                           address-family ipv4 unicast
mpls label range table 0 100000            !
199999                                     !
                                           mpls oam
                                           !
                                           mpls label range table 0 200000
                                           299999
```

其他节点参见上述配置，依次修改 net 地址和 MPLS 标签域。

在完成上述配置后，可以查看各个节点的路由表，例如，对节点 1 上的本地路由、直连路由以及其他路由进行分类查看，可以看到，除了本地路由，节点 1 已经通过 IS-IS 建立了到达所有其他主机和网段的路由。

```
RP/0/0/CPU0:XRV-1#show route ipv4 isis

i L2 2.2.2.2/32 [115/20] via 12.1.1.2, 00:11:38, GigabitEthernet0/0/0/2
i L2 3.3.3.3/32 [115/20] via 13.1.1.3, 00:11:43, GigabitEthernet0/0/0/3
i L2 4.4.4.4/32 [115/30] via 12.1.1.2, 00:11:38, GigabitEthernet0/0/0/2
               [115/30] via 13.1.1.3, 00:11:38, GigabitEthernet0/0/0/3
i L2 5.5.5.5/32 [115/40] via 13.1.1.3, 00:11:43, GigabitEthernet0/0/0/3
i L2 6.6.6.6/32 [115/30] via 13.1.1.3, 00:11:43, GigabitEthernet0/0/0/3
i L2 13.1.1.0/24 [115/20] via 13.1.1.3, 00:11:43, GigabitEthernet0/0/0/3
i L2 24.0.0.0/28 [115/20] via 12.1.1.2, 00:11:38, GigabitEthernet0/0/0/2
i L2 24.1.1.0/28 [115/20] via 12.1.1.2, 00:11:38, GigabitEthernet0/0/0/2
i L2 34.1.1.0/24 [115/20] via 13.1.1.3, 00:11:43, GigabitEthernet0/0/0/3
i L2 45.1.1.0/28 [115/40] via 13.1.1.3, 00:11:43, GigabitEthernet0/0/0/3
i L2 56.1.1.0/28 [115/30] via 13.1.1.3, 00:11:43, GigabitEthernet0/0/0/3
```

6.2.3 协议交互过程

1. SR 能力通告

IS-IS 默认条件下不开启 SR 功能。开启 SR 能力的指令为：
```
router isis N address-family ipv4 unicast segment-routing mpls
```

开启 SR 功能后，IS-IS 生成新的 L2 LSP，其中包含 Router Capability TLV，TLV 中通告 Router ID、SR 算法、SR Range、Node MSD 等信息，IS-IS LSP 报文 Router Capability TLV 格式如图 6.2 所示。

```
∨ Router Capability (t=242, l=70)
    Type: 242
    Length: 70
    Router ID: 0x01010101
    .... ...0 = S bit: False
    .... ..0. = D bit: False
  ∨ Segment Routing - Capability (t=2, l=9)
      1... .... = I flag: IPv4 support: True
      .0.. .... = V flag: IPv6 support: False
      Range: 8000
    ∨ SID/Label (t=1, l=3)
        Label: 16000
  ∨ Node Maximum SID Depth (t=23, l=2)
      MSD Type: Base MPLS Imposition (1)
      MSD Value: 10
  ∨ Segment Routing - Algorithms (t=19, l=4)
      Algorithm: Shortest Path First (SPF) (0)
      Algorithm: Strict Shortest Path First (SPF) (1)
```

图 6.2　IS-IS LSP 报文 Router Capability TLV 格式

Router Capability TLV：用于通告路由器的支持的功能，通过包含不同的子 TLV 通告不同功能，TLV Type 为 242。

Router ID：包含在 Router Capability TLV 里，用于在 SR 域里标识一个路由器。一般是 Loopback 接口 IP 地址。

Segment Routing-Capability Sub-TLV：用于通告 SR-MPLS 的全局 MPLS 标签范围，其中 SID/Label Sub-TLV 包含标签的起始值，此字段和 Range 字段共同决定全局 MPLS 标签的取值范围。Sub-TLV Type 为 2。

SID/Label Sub-TLV：表示一个 MPLS 标签值或 MPLS 标签的索引。Sub-TLV Type 为 1。

Segment Routing-Algorithms Sub-TLV：在计算可达性时，SR 路由器可以使用各种算法，其中 0 为基于度量的 SPF 算法，1 为严格 SPF 算法。Sub-TLV Type 为 19。

Node Maximum SID Depth Sub-TLV：用于通告路由器的最大 SID 栈深，内容由一组 Type 和 Value 组成，该 Value 为路由器所有链接所支持的标签深度最大值。Sub-TLV Type 为 23。

2. AdjSID 通告与转发表的建立

在启动 SR 功能后，节点自动为所有链路进行 AdjSID 的分配和通告。以节点 4 为例进行阐述，它的链路类型比较丰富，既包括点到点链路又包括广播链路。AdjSID 信息采用洪泛方式扩散，可以在任一点查看节点 4 通告的链路状态数据库中 AdjSID 信息，节点 4 记录的 AdjSID 信息如下：

```
RP/0/RP0/CPU0:XRV-4#show isis database 0000.0000.0004.00-00
verbose
//链路1 （GigabitEthernet0/0/0/0）的 AdjSID 信息
  Metric: 10           IS-Extended 0000.0000.0003.01
    Interface IP Address: 34.1.1.4
    Link Maximum SID Depth:
      Label Imposition: 10
    LAN-ADJ-SID: F:0 B:0 V:1 L:1 S:0 P:0 weight:0 Adjacency-sid:
400003 System ID:0000.0000.0006
    LAN-ADJ-SID: F:0 B:0 V:1 L:1 S:0 P:0 weight:0 Adjacency-sid:
400009 System ID:0000.0000.0003
//链路2（GigabitEthernet0/0/0/2）的 AdjSID 信息
  Metric: 10           IS-Extended 0000.0000.0002.00
    Interface IP Address: 24.1.1.4
    Neighbor IP Address: 24.1.1.2
    Link Maximum SID Depth:
      Label Imposition: 10
    ADJ-SID: F:0 B:0 V:1 L:1 S:0 P:0 weight:0 Adjacency-sid:400007
//链路3（GigabitEthernet0/0/0/6）的 AdjSID 信息
  Metric: 10           IS-Extended 0000.0000.0002.00
    Interface IP Address: 24.0.0.4
    Neighbor IP Address: 24.0.0.2
    Link Maximum SID Depth:
      Label Imposition: 10
    ADJ-SID: F:0 B:0 V:1 L:1 S:0 P:0 weight:0 Adjacency-sid:400005
//链路4（GigabitEthernet0/0/0/5）的 AdjSID 信息
  Metric: 10           IS-Extended 0000.0000.0005.00
    Interface IP Address: 45.1.1.4
```

```
Neighbor IP Address: 45.1.1.5
Link Maximum SID Depth:
  Label Imposition: 10
ADJ-SID: F:0 B:0 V:1 L:1 S:0 P:0 weight:0 Adjacency-sid:400001
```

1）链路 1 为 LAN 类型

IS-IS 协议的 LSP 使用 Extended IS Reachability TLV 通告链路上邻接关系信息，在 LAN 链路上，只通告到伪节点 DIS 的邻接关系，TLV Type 为 22。在这个 LAN 上，DIS 为 0000.0000.0003.01，即节点 3。

因为在 LAN 链路上，只通告到伪节点 DIS 的邻接关系，所以对链路上某个实际的邻居，要通告到它的 LAN-AdjSID，必须标明邻居的 System ID，才能建立映射绑定关系。IS-IS 使用 LAN-AdjSID Sub-TLV 包含到某个邻居的 LAN-AdjSID 标签信息和对应邻居的 System ID，还包括表示 LAN-AdjSID 特性的 Flag 信息。Sub-TLV Type 为 32。

在该子网，节点 4 对邻居节点 3（System ID 0000.0000.0003）分配 LAN-AdjSID 400009。因为该子网上还存在另外一个邻居节点 6（System ID 0000.0000.0006），所以针对这个逻辑邻居分配标签 LAN-AdjSID 400003。同时表明本地支持的链路最大标签栈深度为 10。

2）链路 2～链路 4 为点到点类型

点到点链路上只有和对端的邻接关系，只需要通告到对端的 AdjSID，不需要对端的 System ID。IS-IS 使用 AdjSID Sub-TLV，包含点到点链路邻居的 AdjSID 标签信息，还包括表示特性的 Flag 信息。Sub-TLV Type 为 31。

AdjSID 的 Flag 信息里，B Flag 指示该标签是否为备用路径，用于快速重路由功能。AdjSID 一般都含 L Flag，表示本地 MPLS 标签，同时含 V Flag，表示 MPLS 标签值，而不是索引。

链路 2、链路 3 的邻居节点都是节点 2（System ID 0000.0000.0002），但是连接的接口不同，对应两个不同的点到点链路，对链路 2 分配标签 AdjSID 400007，对链路 3 分配标签 AdjSID 400005。

链路 4 的邻居节点是节点 5（System ID 0000.0000.0005），对链路 4 分配标签 AdjSID 400001。

当节点为链路分发 AdjSID 信息后，在本地将产生对应的 MPLS 转发表项。

由于 AdjSID 仅与本地链路有关，因此，所有标签对应的出标签都是 POP 并转发到指定链路的操作，节点 4 关于 AdjSID 的 MPLS 转发表项如下：

```
RP/0/RP0/CPU0:XRV-4#show mpls forwarding
Wed Nov 17 02:12:41.400 UTC
Local    Outgoing   Prefix         Outgoing      Next Hop    Bytes
Label    Label      or ID          Interface                 Switched
------   --------   -----------    ----------    --------    --------
400000   Pop        SR Adj (idx 1) Gi0/0/0/5     45.1.1.5    0
400001   Pop        SR Adj (idx 3) Gi0/0/0/5     45.1.1.5    0
400002   Pop        SR Adj (idx 1) Gi0/0/0/0     34.1.1.6    0
400003   Pop        SR Adj (idx 3) Gi0/0/0/0     34.1.1.6    0
400004   Pop        SR Adj (idx 1) Gi0/0/0/6     24.0.0.2    0
400005   Pop        SR Adj (idx 3) Gi0/0/0/6     24.0.0.2    0
400006   Pop        SR Adj (idx 1) Gi0/0/0/2     24.1.1.2    0
400007   Pop        SR Adj (idx 3) Gi0/0/0/2     24.1.1.2    0
400008   Pop        SR Adj (idx 1) Gi0/0/0/0     34.1.1.3    0
400009   Pop        SR Adj (idx 3) Gi0/0/0/0     34.1.1.3    0
```

可以看出，IS-IS SR 不仅为每个链路生成了一个主 AdjSID，还生成了一个备用 AdjSID，带有 B flag。在有的系统里，备用 Adj SID 也会包含在 LSP 里洪泛，但本书选用的思科 IOS 版本里没有。

3. NodeSID 通告与转发表的建立

为了对全局 SID 进行统一管理，IS-IS 不会自动为节点分配 SID，只能由管理员手动配置。为节点或者前缀分配 SID 的指令为：

```
router isis X interface ifname address-family ipv4 unicast prefix-
sid [strict-spf | algorithm algorithm-number ] {index SID-index |
absolute SID-value} [n-flag-clear] [explicit-null]
```

SID 配置指令在接口模式下，配置的 SID 直接与接口上的 IP 地址关联。

参数 strict-spf 指定其他节点在计算到达该 SID 的路径时，必须使用严格的最短路径优先算法。

参数 algorithm *algorithm-number* 用来指定其他节点在计算到达该 SID 的路径时，使用 algorithm-number 对应的 SR 灵活算法。

SID 配置可以采用绝对值模式或者索引模式，absolute *SID-value* 是直接配置标签值，index *SID-index* 则是配置索引。

默认情况下，IS-IS 会在 PrefixSID 上设置 n-flag，表示它是 NodeSID。对于特定的 PrefixSID（如 Anycast PrefixSID），输入 n-flag-clear 关键字，IS-IS 就不会为该 PrefixSID 设置 n-flag。

参数 explicit-null 用于禁止倒数第二跳弹出，即倒数第二跳节点在进行本地标签与出标签关联绑定时，必须把出标签设置为显式空标签 0。

例如，对节点 1 进行了如下配置：

```
RP/0/RP0/CPU0:XRV-1(config-if)#router isis 1
RP/0/RP0/CPU0:XRV-1(config-isis)#interface Loopback1
RP/0/RP0/CPU0:XRV-1(config-isis-if)#address-family ipv4 unicast
RP/0/RP0/CPU0:XRV-1(config-isis-if-af)# prefix-sid absolute 16005 explicit-null
RP/0/RP0/CPU0:XRV-1(config-isis-if-af)# prefix-sid strict-spf index 55
```

节点 1 通告的 NodeSID 信息，包含 PrefixSID 的 IS-IS LSP 报文格式如图 6.3 所示。从拓展前缀 TLV 的 Flags 字段的 Node 标志可知，该通告为 NodeSID 通告，Node 前缀信息为 1.1.1.1/32。

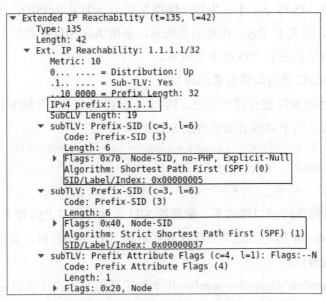

图 6.3 包含 PrefixSID 的 IS-IS LSP 报文格式

IS-IS 使用 Extended IP Reachability TLV 通告前缀路由信息，包含到目的前缀 1.1.1.1/32 的路由信息，TLV Type 为 135。

PrefixSID Sub-TLV：用于通告前缀对应的 SID 索引或 MPLS 标签，包括表示标签特性的 Flag 信息、使用的算法等，Sub-TLV Type 为 3。这里的 N flag 表明此标签是一个 NodeSID。从目的前缀 1.1.1.1/32 的 TLV 中可知，IS-IS 为该前缀分配了两个 SID，一个采用配置 SID 绝对值的值方式，其使用算法 0（普通 SPF 算法）进行路径计算，因为配置了"explicit-null"参数，所以标志位 NP（non-PHP）=E（Explicit-Null）=1；另一个采用配置索引方式，采用算法 1（严格 SPF 算法）进行路径计算，标志位 NP=E=0，无特殊要求。

任意节点收到 NodeSID 信息后，将基于 SRGB+index 机制计算并生成本地 MPLS 转发表项。以节点 2 为例，其针对节点 1 通告的两个 SID 信息，分别计算转发表项。倒数第二跳节点的识别规则是下一跳节点恰恰是通告该 SID 的节点。如果发现自己为倒数第二跳节点，还需要进一步查看源发节点的 NP、E 标志，进而按照源发节点需求的方式进行标签交换。因为节点 2 为倒数第二跳，且目的标签 16005 携带了 NP=1、E=1 标志，所以针对 16005 的出标签为显式空标签，针对 16055 的出标签为隐式空标签。因此，节点 2 的相应的 MPLS 转发表项为：

```
RP/0/RP0/CPU0:XRV-2#show mpls forwarding
Local   Outgoing     Prefix            Outgoing    Next Hop      Bytes
Label   Label        or ID             Interface                 Switched
------  -----------  ----------------  ----------  -----------   --------
16005   Exp-Null-v4  SR Pfx (idx 5)    Gi0/0/0/1   12.1.1.1      0
16055   Pop          SR Pfx (idx 55)   Gi0/0/0/1   12.1.1.1      0
```

4. AnycastSID 通告与转发表的建立

AnycastSID 的配置模式与 NodeSID 一致，但要配置"n-flag-clear"参数，PrefixSID Sub-TLV 的 Flag 中会将 N Flag 设置为 0，表示这并非节点 SID。

为了进行 AnycastSID 通告，需要在相应节点配置一个 Anycast 地址。在节点 2 与节点 3 上配置 Anycast 地址为 23.1.1.23/32，如下：

```
RP/0/RP0/CPU0:XRV-3(config)#interface loopback23 ipv4 address
23.1.1.23 255.255.255.255
```

进入 IS-IS 进程配置 AnycastSID，如下：

```
RP/0/RP0/CPU0:XRV-3(config)#router isis 1
RP/0/RP0/CPU0:XRV-3(config-isis)#interface Loopback23
RP/0/RP0/CPU0:XRV-3(config-isis-if)#address-family ipv4 unicast
RP/0/RP0/CPU0:XRV-3(config-isis-if-af)#prefix-sid index 23 n-flag-clear
```

通过上述配置，节点 2 和节点 3 的环回接口上配置了相同的 IP 地址，并分别为该地址分配标签 16023。网络上其他节点基于各自到节点 2 和节点 3 的最优路径计算下一跳，进而生成关于 AnycastSID 的 MPLS 转发表项。

对于节点 6 而言，它到 Anycast 节点 2 的路径为节点 6→4→2，代价为 20；到节点 3 的路径为节点 6→3，代价为 10。因此，最优下一跳为节点 3，又因为是倒数第二跳，因此，MPLS 转发表项为本地标签 16023，出标签为 POP 操作，下一跳是节点 3 的 IP 34.1.1.3。通过指令查看节点 6 的 MPLS 转发表项为：

```
RP/0/RP0/CPU0:XRV-6#show mpls forwarding
Local   Outgoing   Prefix            Outgoing       Next Hop     Bytes
Label   Label      or ID             Interface                   Switched
------  ---------  ----------------  -------------  -----------  --------
16023   Pop        SR Pfx (idx 23)   Gi0/0/0/0      34.1.1.3     0
```

对于节点 4 而言，它到 Anycast 节点 2 的路径为节点 4→2，代价为 10，下一跳分别为节点 2 的两个不同接口；到 Anycast 节点 3 的路径为节点 4→3，代价为 10，下一跳为节点 3。因此，节点 4 存在 3 条等代价路径到达 AnycastIP，MPLS 转发表项为本地标签 16023，出标签为 POP 操作，通过指令查看节点 4 的 MPLS 转发表项为：

```
RP/0/RP0/CPU0:XRV-4#show mpls forwarding
Local   Outgoing   Prefix            Outgoing       Next Hop     Bytes
Label   Label      or ID             Interface                   Switched
------  ---------  ----------------  -------------  -----------  --------
16023   Pop        SR Pfx (idx 23)   Gi0/0/0/0      34.1.1.3     0
        Pop        SR Pfx (idx 23)   Gi0/0/0/6      24.0.0.2     0
        Pop        SR Pfx (idx 23)   Gi0/0/0/2      24.1.1.2     0
```

对于节点 5 而言，它到节点 2 的路径为节点 5→4→2，到节点 3 的路径为节点 5→4→3 和节点 5→6→3。这意味着节点 5 去往 Anycast 地址存在 3 条等代价路径,但只有两个下一跳。因此,节点 5 将建立起本地标签为 16023,出标签为 16023,且下一跳分别为节点 4 和节点 6 的 MPLS 转发表项,通过指令查看节点 5 的 MPLS 转发表项为:

```
RP/0/RP0/CPU0:XRV-5#show mpls forwarding
Local   Outgoing    Prefix              Outgoing        Next Hop        Bytes
Label   Label       or ID               Interface                       Switched
------  ----------  ------------------  --------------  --------------  ----------
16023   16023       SR Pfx (idx 23)     Gi0/0/0/6       56.1.1.6        0
        16023       SR Pfx (idx 23)     Gi0/0/0/4       45.1.1.4        0
```

节点 5 存在 3 条达到 AnycastIP 的路径,分别为节点 5→4→2、节点 5→4→3 和节点 5→6→3。

IS-IS 协议没有类似 OSPF 的优选规则,如果节点 2 通告 SID 绑定和节点 3 不一致,则 IS-IS 会报错,该 AnycastIP 会不生效,转发表里没有相应表项。正常情况下,节点 5 将标签 16023 的数据包发给节点 4,节点 4 走哪条路径由节点 4 自己决定。通常是基于数据包的相关信息通过轮询方式在多个下一跳节点之间负载均衡。通过指令查看节点 4 的 MPLS 转发表,可知节点 4 到 AnycastSID 16023 有 3 个等价的下一跳,如下:

```
RP/0/RP0/CPU0:XRV-4#show mpls forwarding
Local   Outgoing    Prefix              Outgoing        Next Hop        Bytes
Label   Label       or ID               Interface                       Switched
------  ----------  ------------------  --------------  --------------  ----------
16023   Pop         SR Pfx (idx 23)     Gi0/0/0/0       34.1.1.3        0
        Pop         SR Pfx (idx 23)     Gi0/0/0/6       24.0.0.2        0
        Pop         SR Pfx (idx 23)     Gi0/0/0/2       24.1.1.2        0
```

在节点 5 通过 traceroute 进行路径探测,可知节点 4 通过接口 Gi0/0/0/0 把包转给了节点 3,如下:

```
RP/0/0/CPU0:XRV-5#traceroute sr-mpls 23.1.1.23/32 verbose
  0 45.1.1.5 45.1.1.4 MRU 1500 [Labels: 16023 Exp: 0]
L 1 45.1.1.4 34.1.1.3 MRU 1500 [Labels: implicit-null Exp: 0] 89 ms,
ret code 8
! 2 34.1.1.3 150 ms, ret code 3
```

6.2.4 数据转发验证

1. 多路径转发

SR-MPLS 与传统 MPLS 数据转发的关键不同点是支持 MPLS 的 ECMP，以实现端点之间多条路径资源的充分利用。

当端点之间存在一条路径时，源节点依次发送 TTL 增加的 MPLS Echo 报文，开启 MPLS OAM 服务的节点向源节点回应 TTL 超时报文。由于各节点的下一跳唯一，源节点只需要不断增加发送报文的 TTL 值，并从响应报文中提取回应节点地址即可得到整条路径信息。然而，当节点采用 ECMP 进行数据转发时，各节点去往目的地的下一跳不唯一，它会将流量按照特定哈希规则负载均衡到不同的下一跳节点上。源节点不能仅通过增加 TTL 值来实现路径的发现，它必须携带更加详尽的信息，以指导中间节点将探测数据沿着某个路径发送。中间节点的 TTL 超时报文也必须携带更多的辅助信息来协助源节点完成上述操作。

路径探测原理参见 OSPF 实验相关部分。

下面实验节点 5 探测去往节点 1 的多路径情况。

在节点 1 进行了如下配置：

```
RP/0/RP0/CPU0:XRV-1(config-if)#router isis 1
RP/0/RP0/CPU0:XRV-1(config-isis)#interface Loopback1
RP/0/RP0/CPU0:XRV-1(config-isis-if)#address-family ipv4 unicast
RP/0/RP0/CPU0:XRV-1(config-isis-if-af)# prefix-sid absolute 16005 explicit-null
```

首先查看节点 5 关于 1.1.1.1/32 的 MPLS 转发表信息，如下：

```
RP/0/RP0/CPU0:XRV-5#show mpls forwarding prefix 1.1.1.1/32
Local   Outgoing    Prefix              Outgoing       Next Hop        Bytes
Label   Label       or ID               Interface                      Switched
------  ----------  ------------------  -------------  --------------  ----------
16005   16005       SR Pfx (idx 5)      Gi0/0/0/6      56.1.1.6        0
        16005       SR Pfx (idx 5)      Gi0/0/0/4      45.1.1.4        0
```

从节点 5 的 MPLS 转发表中可以看到，它拥有两个去往目的地 1.1.1.1/32 的下一跳，分别为节点 4 和节点 6。通过 SR-MPLS traceroute 指令进行路径探测，得到的信息如下：
RP/0/RP0/CPU0:XRV-5#traceroute sr-mpls multipath 1.1.1.1/32

```
LL!
Path 0 found,
 output interface GigabitEthernet0/0/0/6 nexthop 56.1.1.6
 source 56.1.1.5 destination 127.0.0.0
LL!
Path 1 found,
 output interface GigabitEthernet0/0/0/4 nexthop 45.1.1.4
 source 45.1.1.5 destination 127.0.0.0
L!
Path 2 found,
 output interface GigabitEthernet0/0/0/4 nexthop 45.1.1.4
 source 45.1.1.5 destination 127.0.0.5
L!
Path 3 found,
 output interface GigabitEthernet0/0/0/4 nexthop 45.1.1.4
 source 45.1.1.5 destination 127.0.0.3

Paths (found/broken/unexplored) (4/0/0)
Echo Request (sent/fail) (10/0)
Echo Reply (received/timeout) (10/0)
Total Time Elapsed 616 ms
```

从上述结果看出，节点 5 共探测发现了 4 条去往目的节点 1 的路径，其中有 3 条路径经过了下一跳节点 4：Path1、Path2、Path3。在探测这 4 条路径时，节点 5 使用了连接节点 4 的接口 IP 地址 45.1.1.5 作为源地址，但是目的地址不同，分别为 127.0.0.0、127.0.0.5、127.0.0.3。在上述过程中，在节点 5 的 Gi0/0/0/4 接口上抓包，得到 SR MPLS traceroute 交互过程，如图 6.4 所示。

下面结合路径探测过程分析节点 5 选用不同 IP 地址进行路径探索的原因。

第一步，节点 5 向节点 4 发送 TTL=1 的 MPLS Echo Request 报文（序号 13），携带多径信息（IP=127.0.0.0，Mask=0xffffffff）。节点 5 请求报文如图 6.5 所示。

No.	Time	Source	Destination	Protocol	Length	Info
13	32.665284329	45.1.1.5	127.0.0.0	MPLS E...	146	MPLS Echo Request
14	32.676587303	45.1.1.4	45.1.1.5	MPLS E...	186	MPLS Echo Reply
15	32.728821866	45.1.1.5	127.0.0.0	MPLS E...	146	MPLS Echo Request
16	32.786297850	34.1.1.3	45.1.1.5	MPLS E...	122	MPLS Echo Reply
17	32.836284630	45.1.1.5	127.0.0.0	MPLS E...	146	MPLS Echo Request
18	32.935481502	13.1.1.1	45.1.1.5	MPLS E...	90	MPLS Echo Reply
19	32.982781990	45.1.1.5	127.0.0.5	MPLS E...	146	MPLS Echo Request
20	33.046120114	24.0.0.2	45.1.1.5	MPLS E...	122	MPLS Echo Reply
21	33.098546224	45.1.1.5	127.0.0.5	MPLS E...	146	MPLS Echo Request
22	33.205110867	12.1.1.1	45.1.1.5	MPLS E...	90	MPLS Echo Reply
23	33.254315995	45.1.1.5	127.0.0.3	MPLS E...	146	MPLS Echo Request
24	33.308937781	24.1.1.2	45.1.1.5	MPLS E...	122	MPLS Echo Reply
25	33.362483074	45.1.1.5	127.0.0.3	MPLS E...	146	MPLS Echo Request
26	33.463843422	12.1.1.1	45.1.1.5	MPLS E...	90	MPLS Echo Reply

图 6.4　SR MPLS traceroute 报文交互过程

```
▼ Downstream Mapping
    Type: Downstream Mapping (2)
    Length: 28
    MTU: 1500
    Address Type: IPv4 Numbered (1)
  ▼ DS Flags: 0x00
      0000 00.. = MBZ: 0x00
      .... ..0. = Interface and Label Stack Request: False
      .... ...0 = Treat as Non-IP Packet: False
    Downstream IP Address: 45.1.1.4
    Downstream Interface Address: 45.1.1.4
    Multipath Type: Bit-masked IPv4 address set (8)
    Depth Limit: 0
    Multipath Length: 8
  ▼ Multipath Information
      IP Address: 127.0.0.0
      Mask: ffffffff
  ▼ Downstream Label Element 1, Label: 16005, Exp: 0, BOS: 1, Protocol: 0 (Unknown)
```

图 6.5　节点 5 请求报文

节点 4 将本地 MPLS 分组转发映射规则通过 DSMAP 反馈给节点 5（序号 14）。节点 4 反馈给节点 5 的 MPLS Echo Reply 报文如图 6.6 所示。

该报文中携带 3 个 DSMAP，说明节点 4 拥有 3 个等价下一跳，从"Downstream IP Address"获知它们是 34.1.1.3、24.0.0.2 和 24.1.1.2，从"Downstream Label Element"域得知出标签为 16005。"Multipath Information"则表明节点 4 在 3 个下一跳之间的 MPLS 分组转发规则，其携带"IP Address"（初始 IP）和"Mask"两个字段。前者可理解为"初始 IP 地址"，由探测节点 5 发出 MPLS Echo Request 时设定；后者可理解为"地址增量"，指明当前节点把自给定初始 IP 开始的连续 32 个地址中的哪些绑定到相应下一跳上。Mask 长度 4byte，每一个比特位对应初始 IP 地址。例如，节点 4 反馈的第一条多径信息 Mask 为 0xe2230013，二进制数为"1110 0010 0010 0011 0000 0000 0001 0011"，按照从左到右顺序，置 1 的为比特位数分别为第 0、1、2、6、10、14、15、

27、30、31。IP 地址组内的 IP 地址计算方法为：初始 IP 地址+比特位数，即 IPS1={127.0.0.0、127.0.0.1、127.0.0.2、127.0.0.6、127.0.0.10、127.0.0.14、127.0.0.15、127.0.0.27、127.0.0.30、127.0.0.31}。通过上述分析可知，目的 IP 属于 IPS1 的探测分组将被节点 4 转发给 34.1.1.3。

```
▶ Vendor Private
▼ Downstream Mapping
    Type: Downstream Mapping (2)
    Length: 28
    MTU: 1500
    Address Type: IPv4 Numbered (1)
  ▶ DS Flags: 0x00
    Downstream IP Address: 34.1.1.3
    Downstream Interface Address: 34.1.1.3
    Multipath Type: Bit-masked IPv4 address set (8)
    Depth Limit: 0
    Multipath Length: 8
  ▼ Multipath Information
      IP Address: 127.0.0.0
      Mask: e2230013
  ▶ Downstream Label Element 1, Label: 16005, Exp: 0, BOS: 1, Protocol: 0 (Unknown)
▼ Downstream Mapping
    Type: Downstream Mapping (2)
    Length: 28
    MTU: 1500
    Address Type: IPv4 Numbered (1)
  ▶ DS Flags: 0x00
    Downstream IP Address: 24.0.0.2
    Downstream Interface Address: 24.0.0.2
    Multipath Type: Bit-masked IPv4 address set (8)
    Depth Limit: 0
    Multipath Length: 8
  ▼ Multipath Information
      IP Address: 127.0.0.0
      Mask: 05443ea4
  ▶ Downstream Label Element 1, Label: 16005, Exp: 0, BOS: 1, Protocol: 0 (Unknown)
▼ Downstream Mapping
    Type: Downstream Mapping (2)
    Length: 28
    MTU: 1500
    Address Type: IPv4 Numbered (1)
  ▶ DS Flags: 0x00
    Downstream IP Address: 24.1.1.2
    Downstream Interface Address: 24.1.1.2
    Multipath Type: Bit-masked IPv4 address set (8)
    Depth Limit: 0
    Multipath Length: 8
  ▼ Multipath Information
      IP Address: 127.0.0.0
      Mask: 1898c148
  ▶ Downstream Label Element 1, Label: 16005, Exp: 0, BOS: 1, Protocol: 0 (Unknown)
```

图 6.6　节点 4 反馈给节点 5 的 MPLS Echo Reply 报文

采用相同的方法分析节点 4 的第二条路径，Mask=0x05443ea4=0000 0101 0100 0100 0011 1110 1010 0100，所以，IPS2={127.0.0.5、127.0.0.7、127.0.0.9、127.0.0.13、127.0.0.18、127.0.0.19、127.0.0.20、127.0.0.21、127.0.0.22、127.0.0.24、127.0.0.26、127.0.0.29}。目的 IP 属于 IPS2 的探测分组将被节点 2 转发给 24.0.0.2。

节点 4 的第三条路径，Mask=0x1898c148=0001 1000 1001 1000 1100 0001 0100 1000，所以，IPS3={127.0.0.3、127.0.0.4、127.0.0.8、127.0.0.11、127.0.0.12、127.0.0.16、127.0.0.17、127.0.0.23、127.0.0.25、127.0.0.28}。目的 IP 属于 IPS3 的探测分组将被节点 2 转发给 24.1.1.2。

也可以使用 ping 指令查询路径上某个节点的 DSMAP 信息，例如，节点 5 通过设置 TTL=1 并携带 DSMAP 参数，可以让节点 4 反馈其 DSMAP 信息，指令与结果如下：

```
RP/0/0/CPU0:XRV-1#ping sr-mpls 1.1.1.1/32 dsmap ttl 1 output nexthop
45.1.1.4 repeat 1


L Echo Reply received from 45.1.1.4
  DSMAP 0, DS Router Addr 34.1.1.3, DS Intf Addr 34.1.1.3
    Depth Limit 0, MRU 1500 [Labels: 16005 Exp: 0]
    Multipath Addresses:
      127.0.0.1      127.0.0.2      127.0.0.6      127.0.0.10
      127.0.0.14     127.0.0.15     127.0.0.27     127.0.0.30
      127.0.0.31
  DSMAP 1, DS Router Addr 24.0.0.2, DS Intf Addr 24.0.0.2
    Depth Limit 0, MRU 1500 [Labels: 16005 Exp: 0]
    Multipath Addresses:
      127.0.0.5      127.0.0.7      127.0.0.9      127.0.0.13
      127.0.0.18     127.0.0.19     127.0.0.20     127.0.0.21
      127.0.0.22     127.0.0.24     127.0.0.26     127.0.0.29
      127.0.0.32
  DSMAP 2, DS Router Addr 24.1.1.2, DS Intf Addr 24.1.1.2
    Depth Limit 0, MRU 1500 [Labels: 16005 Exp: 0]
    Multipath Addresses:
      127.0.0.3      127.0.0.4      127.0.0.8      127.0.0.11
      127.0.0.12     127.0.0.16     127.0.0.17     27.0.0.23
      127.0.0.25     127.0.0.28
```

依据节点 4 反馈的下游映射信息，节点 5 可获知，采用 IPS1 内的地址作为探测分组目的 IP 地址，则节点 4 会将其转发给下一跳 34.1.1.3；采用 IPS2 内的地址作为探测分组目的 IP 地址，则节点 4 会将其转发给下一跳 24.0.0.2，采用 IPS3 内的地址作为探测分组目的 IP 地址，则节点 4 会将其转发给下一跳 24.1.1.2。

第二步，节点 5 对应节点 4 的第一个路径 IPS1 的下一跳 34.1.1.3 发送 MPLS Echo Request 报文（序号 15），携带多径信息（IP=127.0.0.0，Mask= 0xe2230013），如图 6.7 所示。

```
▼ Downstream Mapping
    Type: Downstream Mapping (2)
    Length: 28
    MTU: 1500
    Address Type: IPv4 Numbered (1)
  ▼ DS Flags: 0x00
      0000 00.. = MBZ: 0x00
      .... ..0. = Interface and Label Stack Request: False
      .... ...0 = Treat as Non-IP Packet: False
    Downstream IP Address: 34.1.1.3
    Downstream Interface Address: 34.1.1.3
    Multipath Type: Bit-masked IPv4 address set (8)
    Depth Limit: 0
    Multipath Length: 8
  ▼ Multipath Information
      IP Address: 127.0.0.0
      Mask: e2230013
  ▶ Downstream Label Element 1, Label: 16005, Exp: 0, BOS: 1, Protocol: 0 (Unknown)
```

图 6.7　节点 5 请求报文

节点 3 将本地 MPLS 分组转发映射规则通过 DSMAP 反馈给节点 5（序号 16），如图 6.8 所示。

```
▼ Downstream Mapping
    Type: Downstream Mapping (2)
    Length: 28
    MTU: 1500
    Address Type: IPv4 Numbered (1)
  ▼ DS Flags: 0x00
      0000 00.. = MBZ: 0x00
      .... ..0. = Interface and Label Stack Request: False
      .... ...0 = Treat as Non-IP Packet: False
    Downstream IP Address: 13.1.1.1
    Downstream Interface Address: 13.1.1.1
    Multipath Type: Bit-masked IPv4 address set (8)
    Depth Limit: 0
    Multipath Length: 8
  ▼ Multipath Information
      IP Address: 127.0.0.0
      Mask: e2230013
  ▶ Downstream Label Element 1, Label: 0 (IPv4 Explicit-Null), Exp: 0, BOS: 1, Protocol: 0 (Unknown)
```

图 6.8　节点 3 回应报文

按照前述规则，节点 3 只有一个下一跳，Mask 为 0xe2230013，IPS1 等同于节点 4 的 IPS1，转发给下一跳节点 13.1.1.1。从 "Downstream Label Element" 域得知出标签为 NULL，这是因为节点 3 已经是倒数第二跳。

第三步，节点 5 向 13.1.1.1 发送的 MPLS Echo Request 报文（序号 17）携带多径信息（IP=127.0.0.0，Mask=0xe2230013），如图 6.9 所示。

```
▼ Downstream Mapping
    Type: Downstream Mapping (2)
    Length: 28
    MTU: 1500
    Address Type: IPv4 Numbered (1)
  ▶ DS Flags: 0x00
    Downstream IP Address: 13.1.1.1
    Downstream Interface Address: 13.1.1.1
    Multipath Type: Bit-masked IPv4 address set (8)
    Depth Limit: 0
    Multipath Length: 8
  ▼ Multipath Information
      IP Address: 127.0.0.0
      Mask: e2230013
  ▶ Downstream Label Element 1, Label: 0 (IPv4 Explicit-Null), Exp: 0, BOS: 1, Protocol: 0 (Unknown)
```

图 6.9　节点 5 请求报文

节点 1 在 13.1.1.1 接口收到该分组后，使用该接口地址进行回应，因为它是探测的目标节点，因此，报文（序号 18）中不再携带 DSMAP 信息。

至此，节点 1 已经完成对 Path0 路径的探测，它依次经过节点 4、节点 3 到达节点 1。

对于 Path1，第一步，节点 5 对应节点 4 的第二个路径 IPS2 的下一跳 24.0.0.2 发送的 MPLS Echo Request 报文（序号 19），携带多径信息（IP=127.0.0.0，Mask=0x05443ea4）。

节点 2 将本地 MPLS 分组转发映射规则通过 DSMAP 反馈给节点 5（序号 20），如图 6.10 所示。节点 2 只有一个下一跳，Mask 为 0x05443ea4，转发给下一跳节点 12.1.1.1。从 "Downstream Label Element" 域得知出标签为 NULL，这是因为节点 2 已经是倒数第二跳。

```
▼ Downstream Mapping
    Type: Downstream Mapping (2)
    Length: 28
    MTU: 1500
    Address Type: IPv4 Numbered (1)
  ▼ DS Flags: 0x00
      0000 00.. = MBZ: 0x00
      .... ..0. = Interface and Label Stack Request: False
      .... ...0 = Treat as Non-IP Packet: False
    Downstream IP Address: 12.1.1.1
    Downstream Interface Address: 12.1.1.1
    Multipath Type: Bit-masked IPv4 address set (8)
    Depth Limit: 0
    Multipath Length: 8
  ▼ Multipath Information
      IP Address: 127.0.0.0
      Mask: 05443ea4
  ▶ Downstream Label Element 1, Label: 0 (IPv4 Explicit-Null), Exp: 0, BOS: 1, Protocol: 0 (Unknown)
```

图 6.10　节点 2 回应报文

第二步，节点 5 向 12.1.1.1 发送的 MPLS Echo Request 报文（序号 21），携带多径信息（IP=127.0.0.0，Mask=0x05443ea4）。

节点 1 在 12.1.1.1 接口收到该分组后，使用该接口地址进行回应，因为它是探测的目标节点，因此，报文（序号 22）中不再携带 DSMAP 信息

至此，节点 5 已经完成对 Path1 路径的探测，它依次经过节点 4、24.1.1.4—24.1.1.2 链路、节点 2 最后到达节点 1。

采用类似的方式，节点 5 探测得到 Path2 和 Path3。

- Path2：节点 4、24.0.0.4—24.0.0.2 链路、节点 2、节点 1（序号 23、序号 24）。
- Path3：节点 6、节点 3、节点 1（序号 25、序号 26）。

详细路径探测信息如下：

```
RP/0/0/CPU0:XRV-5#traceroute sr-mpls multipath 1.1.1.1/32 verbose

LL!
Path 0 found,
 output interface GigabitEthernet0/0/0/4 nexthop 45.1.1.4
 source 45.1.1.5 destination 127.0.0.0
  0 45.1.1.5 45.1.1.4 MRU 1500 [Labels: 16005 Exp: 0] multipaths 0
L 1 45.1.1.4 34.1.1.3 MRU 1500 [Labels: 16005 Exp: 0] ret code 8
multipaths 3
L 2 34.1.1.3 13.1.1.1 MRU 1500 [Labels: explicit-null Exp: 0] ret code 8
multipaths 1
! 3 13.1.1.1, ret code 3 multipaths 0
L!
Path 1 found,
 output interface GigabitEthernet0/0/0/4 nexthop 45.1.1.4
 source 45.1.1.5 destination 127.0.0.3
  0 45.1.1.5 45.1.1.4 MRU 1500 [Labels: 16005 Exp: 0] multipaths 0
L 1 45.1.1.4 24.1.1.2 MRU 1500 [Labels: 16005 Exp: 0] ret code 8
multipaths 3
L 2 24.1.1.2 12.1.1.1 MRU 1500 [Labels: explicit-null Exp: 0] ret code 8
multipaths 1
! 3 12.1.1.1, ret code 3 multipaths 0
```

```
Path 2 found,
 output interface GigabitEthernet0/0/0/4 nexthop 45.1.1.4
 source 45.1.1.5 destination 127.0.0.5
   0 45.1.1.5 45.1.1.4 MRU 1500 [Labels: 16005 Exp: 0] multipaths 0
 L 1 45.1.1.4 24.0.0.2 MRU 1500 [Labels: 16005 Exp: 0] ret code 8
multipaths 3
 L 2 24.0.0.2 12.1.1.1 MRU 1500 [Labels: explicit-null Exp: 0] ret code 8
multipaths 1
 ! 3 12.1.1.1, ret code 3 multipaths 0
 L!
Path 3 found,
 output interface GigabitEthernet0/0/0/6 nexthop 56.1.1.6
 source 56.1.1.5 destination 127.0.0.0
   0 56.1.1.5 56.1.1.6 MRU 1500 [Labels: 16005 Exp: 0] multipaths 0
 L 1 56.1.1.6 34.1.1.3 MRU 1500 [Labels: 16005 Exp: 0] ret code 8
multipaths 1
 L 2 34.1.1.3 13.1.1.1 MRU 1500 [Labels: explicit-null Exp: 0] ret code 8
multipaths 1
 ! 3 13.1.1.1, ret code 3 multipaths 0
 LL!

Paths (found/broken/unexplored) (4/0/0)
 Echo Request (sent/fail) (10/0)
 Echo Reply (received/timeout) (10/0)
 Total Time Elapsed 1 seconds 575 ms
```

2. 任意路径转发

SR-MPLS 与传统 MPLS 数据转发的另一个不同点是支持任意路径编排控制。

实验开始前，配置节点 N 的 Loopback1 的 NodeSID 为 $1600N$，例如，在节点 1 进行了如下配置：

```
RP/0/RP0/CPU0:XRV-1(config-if)#router isis 1
RP/0/RP0/CPU0:XRV-1(config-isis)#interface Loopback1
RP/0/RP0/CPU0:XRV-1(config-isis-if)#address-family ipv4 unicast
```

```
RP/0/RP0/CPU0:XRV-1(config-isis-if-af)# prefix-sid absolute 16001
```

节点 1 去往目的节点 6 的默认路径为节点 1→3→6，通过 SR-MPLS traceroute 进行探测，得到结果如下：

```
RP/0/RP0/CPU0:XRV-1#traceroute sr-mpls 6.6.6.6/32 verbose

  0 13.1.1.1 13.1.1.3 MRU 1500 [Labels: 16006 Exp: 0]
L 1 13.1.1.3 34.1.1.6 MRU 1500 [Labels: implicit-null Exp: 0] 43 ms,
ret code 8
! 2 34.1.1.6 25 ms, ret code 3
```

如果期望节点1去往节点6的数据按照节点1→3→4→2→4→5→6路径进行转发，可通过NodeSID进行路径编排。例如，节点2发给节点1的MPLS分组头部依次压入代表路径上各节点的标签，从栈顶到栈底依次为：16003,16004,16002,16004,16005,16006。由于采用了自编排路径，因此，要使用SR-MPLS的Nil-FEC进行路径测试与验证，探测指令与结果如下：

```
RP/0/0/CPU0:XRV-2#traceroute sr-mpls nil-fec labels 16003, 16004, 16002,
16004, 16005, 16006 output interface g0/0/0/1 nexthop 12.1.1.1 verbose
  0 12.1.1.2 12.1.1.1 MRU 1500 [Labels: 16003/16004/16002/16004/16005/
16006/explicit-null Exp: 0/0/0/0/0/0/0]
L 1 12.1.1.1 13.1.1.3 MRU 1500 [Labels: implicit-null/16004/16002/16004/
16005/16006/explicit-null Exp: 0/0/0/0/0/0/0] 65 ms, ret code 8
L 2 13.1.1.3 34.1.1.4 MRU 1500 [Labels: implicit-null/16002/16004/
16005/16006/explicit-null Exp: 0/0/0/0/0/0] 118 ms, ret code 8
L 3 34.1.1.4 24.1.1.2 MRU 1500 [Labels: implicit-null/16004/16005/16006/
explicit-null Exp: 0/0/0/0/0] 61 ms, ret code 8
L 4 24.1.1.2 24.1.1.4 MRU 1500 [Labels: implicit-null/16005/16006/
explicit-null Exp: 0/0/0/0] 65 ms, ret code 8
L 5 24.1.1.4 45.1.1.5 MRU 1500 [Labels: implicit-null/16006/explicit-
null Exp: 0/0/0] 58 ms, ret code 8
L 6 45.1.1.5 56.1.1.6 MRU 1500 [Labels: implicit-null/explicit-null
Exp: 0/0] 55 ms, ret code 8
! 7 56.1.1.6 114 ms, ret code 3
```

除了使用NodeSID进行路径编排实现流量引导，也可以采用AdjSID进行流量引导。可以查看每个节点为其直连链路分配的标签信息，假定节点 3 发给节点 1 的 MPLS 分组期望流量依次经过链路 12.1.1.1→12.1.1.2, 24.1.1.2→ 24.1.1.4, 45.1.1.4→45.1.1.5, 56.1.1.5→56.1.1.6。那么可以使用标签栈信息 {100001, 200005, 400001, 500003}，探测指令与结果如下：

```
RP/0/0/CPU0:XRV-2#traceroute sr-mpls nil-fec labels 100001,200005,
400001,500003 output interface g0/0/0/1 nexthop 12.1.1.1 verbose

0 13.1.1.3 13.1.1.1 MRU 1500 [Labels: 100001/200005/400001/500003/
explicit-null Exp: 0/0/0/0/0]
L 1 13.1.1.1 12.1.1.2 MRU 1500 [Labels: implicit-null/200005/
400001/500003/ explicit-null Exp: 0/0/0/0/0] 47 ms, ret code 8
L 2 12.1.1.2 24.1.1.4 MRU 1500 [Labels: implicit-null/400001/
500003/explicit-null Exp: 0/0/0/0] 21 ms, ret code 8
L 3 24.1.1.4 45.1.1.5 MRU 1500 [Labels: implicit-null/500003/
explicit-null Exp: 0/0/0] 27 ms, ret code 8
L 4 45.1.1.5 56.1.1.6 MRU 1500 [Labels: implicit-null/explicit-null
Exp: 0/0] 65 ms, ret code 8
! 5 56.1.1.6 69 ms, ret code 3
```

单纯使用 NodeSID 或者 AdjSID 实现路径编排，往往会使得标签栈过深，而一些硬件设备支持的标签栈深度是有限的，因此，在实际应用中应当将两者进行组合使用，以更为精简的标签栈实现同样的流量引导效果。例如，节点 3 发给节点 1 的 MPLS 分组期望采用上例中基于 AdjSID 编排的路径，采用 NodeSID 和 AdjSID 混编为 {16002,200005,400001,16006}，探测指令与结果如下：

```
RP/0/0/CPU0:XRV-2#traceroute sr-mpls nil-fec labels 16002, 200005,
 400001,16006 output interface g0/0/0/1 nexthop 13.1.1.1 verbose

0 13.1.1.3 13.1.1.1 MRU 1500 [Labels: 16002/200005/400001/16006/
explicit-null Exp: 0/0/0/0/0]
L 1 13.1.1.1 12.1.1.2 MRU 1500 [Labels: implicit-null/200005/
400001/16006/explicit-null Exp: 0/0/0/0/0] 63 ms, ret code 8
L 2 12.1.1.2 24.1.1.4 MRU 1500 [Labels: implicit-null/400001/
16006/explicit-null Exp: 0/0/0/0] 160 ms, ret code 8
L 3 24.1.1.4 45.1.1.5 MRU 1500 [Labels: implicit-null/16006/explicit-
null Exp: 0/0/0] 189 ms, ret code 8
```

```
L 4 45.1.1.5 56.1.1.6 MRU 1500 [Labels: implicit-null/explicit-null
Exp: 0/0] 161 ms, ret code 8
! 5 56.1.1.6 180 ms, ret code 3
```

6.3 域间 SR 实验

区域间相互通告前缀信息，相应 PrefixSID 也伴随着在区域之间洪泛。域间 SR 实验主要围绕 PrefixSID 的跨区域传播展开，演示 IS-IS 协议在多个区域中传播 SID 信息的过程。同时验证在多个区域中，利用 SR 标签进行数据转发的能力。

6.3.1 实验环境

多区域 SR 实验拓扑如图 6.11 所示，所有路由器运行了 IS-IS 协议，节点 1～节点 6 处于同一个 L2 区域中，拓扑同域内实验部分。节点 1、节点 5、节点 6 配置成区域边界 L1/2 路由器。节点 2～节点 4 是纯 L2 路由器。新增加了 3 个纯 L1 路由器，即节点 7、节点 8、节点 9。节点 7、节点 1 同属于一个 L1 区域 1，节点 8、节点 9、节点 5、节点 6 同属于另一个 L1 区域 2。

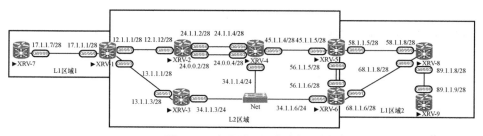

图 6.11 多区域 SR 实验拓扑

所有 IS-IS 链路代价均为默认的 10。在节点 N 设置环回接口，并配置 IP 地址 $N.N.N.N/32$。各节点通过 IS-IS 协议交互网络拓扑信息，建立到所有接口网段和主机地址的路由。

6.3.2 协议配置

按照实验要求，在每个节点配置 IS-IS 协议，在各互联接口、本地环回接口上启动 IS-IS。

在节点 N 开启 SR 功能，并给 Loopback1 接口配置 NodeSID index N。

节点 1～节点 6 的 SRGB 为默认的[16000, 23999]，节点 7、节点 8、节点 9 的 SRGB 分别为[70000, 71000]、[80000, 81000]、[90000, 91000]。

SR 在为本地路由分配标签时，要从本地标签池中申请标签，为了简化，节点 N 的本地标签池为 N00000～N99999。节点协议配置信息如下：

```
节点1：
router isis 1
 net 01.0000.0000.0001.00
 address-family ipv4 unicast
  metric-style wide
  segment-routing mpls
 !
 interface Loopback1
  address-family ipv4 unicast
   prefix-sid index 1
  !
 !
 interface GigabitEthernet0/0/0/0
  point-to-point
  address-family ipv4 unicast
  !
 !
 interface GigabitEthernet0/0/0/2
  point-to-point
  address-family ipv4 unicast
  !
 !
 interface GigabitEthernet0/0/0/3
```

```
  point-to-point
  address-family ipv4 unicast
  !
 !
!
mpls oam
!
mpls label range table 0 100000 199999
```

节点 2：

```
router isis 1
 is-type level-2-only
 net 10.0000.0000.0002.00
 address-family ipv4 unicast
  metric-style wide
  segment-routing mpls
 !
 interface Loopback1
  address-family ipv4 unicast
   prefix-sid index 2
  !
 !
 interface GigabitEthernet0/0/0/1
  point-to-point
  address-family ipv4 unicast
  !
 !
 interface GigabitEthernet0/0/0/4
  point-to-point
  address-family ipv4 unicast
  !
 !
 interface GigabitEthernet0/0/0/6
  point-to-point
```

```
 address-family ipv4 unicast
 !
 !
!
mpls oam
!
mpls label range table 0 200000 299999
```

节点 7：
```
interface Loopback1
 ipv4 address 7.7.7.7 255.255.255.255
!
interface GigabitEthernet0/0/0/0
 ipv4 address 17.1.1.7 255.255.255.240
!
interface GigabitEthernet0/0/0/0
 ipv4 address 17.1.1.7 255.255.255.240
!
router isis 1
 is-type level-1
 net 01.0000.0000.0007.00
 segment-routing global-block 70000 71000
 address-family ipv4 unicast
  metric-style wide
  segment-routing mpls
 !
 interface Loopback1
  address-family ipv4 unicast
   prefix-sid index 7
  !
 !
 interface GigabitEthernet0/0/0/0
  point-to-point
  address-family ipv4 unicast
```

```
!
!
!
mpls oam
!
mpls label range table 0 700000 799999
```

节点 3 和节点 4 的配置参见节点 2，并修改 net 地址和 MPLS 标签域，区域号设为 10。节点 5 和节点 6 的配置参见节点 1，并修改 net 地址和 MPLS 标签域，区域号设为 02。节点 8 和节点 9 的配置参见节点 7，并修改 net 地址和 MPLS 标签域，区域号设为 02。

6.3.3 协议交互

在 IS-IS 协议设计中，2 个 L1 区域之间通常不能直接通信，必须经过 L2 骨干网。通常区域边界 L1/2 路由器会通过设置 ATT 位，让 L1 路由器生成到外界的默认路由。

1. L1 路由汇入 L2

节点 7 会向节点 1 发送自己的 L1 LSP，通告自己的 L1 路由。

IS-IS 默认会把 L1 的路由信息，包括 SR 标签自动汇入 L2 层路由。在节点 1 查看 IS-IS 路由数据库的详细信息，如下：

```
RP/0/RP0/CPU0:XRV-1#show isis d level 2 XRV-1.00-00 verbose
IS-IS 1 (Level-2) Link State Database
LSPID           LSP Seq Num   LSP Checksum   LSP Holdtime/Rcvd    ATT/P/OL
XRV-1.00-00  * 0x00000024    0x3b85         815       /*          0/0/0
  Area Address:  01
  NLPID:         0xcc
  IP Address:    1.1.1.1
  Metric: 20           IP-Extended 7.7.7.7/32
    Prefix-SID Index: 7, Algorithm:0, R:1 N:1 P:1 E:0 V:0 L:0
    Prefix Attribute Flags: X:0 R:1 N:1
  Metric: 10           IP-Extended 12.1.1.0/28
    Prefix Attribute Flags: X:0 R:0 N:0
```

```
Metric: 10            IP-Extended 13.1.1.0/28
  Prefix Attribute Flags: X:0 R:0 N:0
Metric: 10            IP-Extended 17.1.1.0/28
  Prefix Attribute Flags: X:0 R:0 N:0
Metric: 10            IP-Extended 1.1.1.1/32
  Prefix-SID Index: 1, Algorithm:0, R:0 N:1 P:0 E:0 V:0 L:0
  Prefix Attribute Flags: X:0 R:0 N:1
```

可以看到，节点 7 的 L1 的路由 7.7.7.7/32，index 7 已经自动汇入节点 1 的 L2 LSP 中。注意 index 7 包含 R-Flag，表示该 SID 是从其他 level 或其他协议导入的。

类似在节点 5 查看 IS-IS 链路状态数据库（Link State DataBase，LSDB）的详细信息，如下：

```
RP/0/RP0/CPU0:XRV-5#show isis d level 2 XRV-5.00-00 verbose

IS-IS 1 (Level-2) Link State Database
LSPID            LSP Seq Num   LSP Checksum   LSP Holdtime/Rcvd   ATT/P/OL
XRV-5.00-00  * 0x0000001d    0x3f74          925    /*           0/0/0
  Area Address:    02
  NLPID:           0xcc
  IP Address:      5.5.5.5
  Metric: 10            IP-Extended 5.5.5.5/32
    Prefix-SID Index: 5, Algorithm:0, R:0 N:1 P:0 E:0 V:0 L:0
    Prefix Attribute Flags: X:0 R:0 N:1
  Metric: 10            IP-Extended 45.1.1.0/28
    Prefix Attribute Flags: X:0 R:0 N:0
  Metric: 10            IP-Extended 56.1.1.0/28
    Prefix Attribute Flags: X:0 R:0 N:0
  Metric: 10            IP-Extended 58.1.1.0/28
    Prefix Attribute Flags: X:0 R:0 N:0
  Metric: 20            IP-Extended 8.8.8.8/32
    Prefix-SID Index: 8, Algorithm:0, R:1 N:1 P:1 E:0 V:0 L:0
    Prefix Attribute Flags: X:0 R:1 N:1
  Metric: 20            IP-Extended 68.1.1.0/28
    Prefix Attribute Flags: X:0 R:1 N:0
```

```
  Metric: 20          IP-Extended 89.1.1.0/28
    Prefix Attribute Flags: X:0 R:1 N:0
  Metric: 30          IP-Extended 9.9.9.9/32
    Prefix-SID Index: 9, Algorithm:0, R:1 N:1 P:1 E:0 V:0 L:0
    Prefix Attribute Flags: X:0 R:1 N:1
```

可以看到，节点 8 的 L1 的路由 8.8.8.8/32，index 8；节点 9 的 L1 的路由 9.9.9.9/32，index 9 已经自动汇入节点 5 的 L2 LSP 中。

注意到上述 L1 index 汇入 L2 时，会将 NP 标志位（P 标志位，即 no-PHP）置 1，禁止倒数第二跳提前弹出标签。这是由于 L1/2 路由器通告本区域的 L1 index 时，是在自己的 L2 LSP 里通告的，并没有包含这条 index 的来源。这使得 L2 域的其他节点在判定倒数第二跳时产生了假象。例如，节点 1 通告 7.7.7.7 的 index 时将通告路由器设定为自己，使得节点 2 认为节点 1 直连 7.7.7.7，于是认定自己为去往 7.7.7.7 的倒数第二跳，并按照 NP=0 且 E=0 标志位安装标签弹出表项，这使得段标签在没有进入目的区域时就被弹出了，无法达到 SID 引导流量的目的。因此，L1/2 路由器在中继 PrefixSID 时将 NP 比特置 1，确保倒数第二跳节点不执行标签弹出操作，而执行标签交换操作，从而保留了 SID 标签信息，这样后续转发能够继续按照 SID 引导流量。

在节点 2 上查看 MPLS 转发表，可见在转发标签 index7 时，不会弹出该标签。如下：

```
RP/0/RP0/CPU0:XRV-2#show mpls forwarding
Local   Outgoing    Prefix              Outgoing        Next Hop      Bytes
Label   Label       or ID               Interface                     Switched
------  ----------  ------------------  --------------  ------------  --------
16001   Pop         SR Pfx (idx 1)      Gi0/0/0/1       12.1.1.1      0
16003   16003       SR Pfx (idx 3)      Gi0/0/0/6       24.0.0.4      0
        16003       SR Pfx (idx 3)      Gi0/0/0/4       24.1.1.4      0
        16003       SR Pfx (idx 3)      Gi0/0/0/1       12.1.1.1      0
16004   Pop         SR Pfx (idx 4)      Gi0/0/0/6       24.0.0.4      0
        Pop         SR Pfx (idx 4)      Gi0/0/0/4       24.1.1.4      0
16005   16005       SR Pfx (idx 5)      Gi0/0/0/6       24.0.0.4      0
        16005       SR Pfx (idx 5)      Gi0/0/0/4       24.1.1.4      0
```

16006	16006	SR Pfx (idx 6)	Gi0/0/0/6	24.0.0.4	0
	16006	SR Pfx (idx 6)	Gi0/0/0/4	24.1.1.4	0
16007	16007	SR Pfx (idx 7)	Gi0/0/0/1	12.1.1.1	0

但同时 L2 的路由信息，包括 SR 标签，并不会自动汇入 L1 层路由。

在节点 7 上查看 MPLS 转发表，可见节点 7 只有到 L1 域内节点 1 的 MPLS 转发表项，不存在到其他节点的 MPLS 转发表项。如下：

```
RP/0/RP0/CPU0:XRV-7#show mpls forwarding
Local  Outgoing    Prefix              Outgoin     Next Hop    Bytes
Label  Label       or ID               Interface               Switched
------ ----------- ------------------- ----------- ----------- --------
70001  Pop         SR Pfx (idx 1)      Gi0/0/0/0   17.1.1.1    0
       16006       SR Pfx (idx 6)      Gi0/0/0/4   24.1.1.4    0
```

查看节点 7 路由表，可以看到节点 7 有一条默认路由到节点 1，所有到 L1 区域外的流量都由节点 1 转发。如下：

```
RP/0/RP0/CPU0:XRV-7#show route ipv4

i*L1 0.0.0.0/0 [115/10] via 17.1.1.1, 04:18:47, GigabitEthernet0/0/0/0
i L1 1.1.1.1/32 [115/20] via 17.1.1.1, 04:32:49, GigabitEthernet0/0/0/0
L    7.7.7.7/32 is directly connected, 05:32:18, Loopback1
i L1 12.1.1.0/28 [115/20] via 17.1.1.1, 04:40:27, GigabitEthernet0/0/0/0
i L1 13.1.1.0/28 [115/20] via 17.1.1.1, 04:40:27, GigabitEthernet0/0/0/0
C    17.1.1.0/28 is directly connected, 05:31:31, GigabitEthernet0/0/0/0
L    17.1.1.7/32 is directly connected, 05:31:31, GigabitEthernet0/0/0/0
L    127.0.0.0/8 [0/0] via 0.0.0.0, 05:32:20
```

由此可见，一般情况下 L2 路由器可以知道全网的 index 信息，但 L1 路由器只能知道本区域的 index 信息，无法知道其他区域的 index 信息。

2. L2 路由汇入 L1

在某些情况下，L1/2 路由器也可以把 L2 路由信息注入 L1 路由，优化 L1 路由选择。

在节点 1 里通过命令把 L2 路由信息注入 L1 区域 1，如下：

```
RP/0/RP0/CPU0:XRV-1(config)#route-policy 1
RP/0/RP0/CPU0:XRV-1(config-rpl)#done
```

```
RP/0/RP0/CPU0:XRV-1(config)#router isis 1
RP/0/RP0/CPU0:XRV-1(config-isis)#address-family ipv4 unicast
RP/0/RP0/CPU0:XRV-1(config-isis-af)#propagate level 2 into level 1
route-policy 1
```

在节点 5 里通过命令把 L2 路由信息注入 L1 区域 2,如下:

```
RP/0/RP0/CPU0:XRV-5(config)#route-policy 1
RP/0/RP0/CPU0:XRV-5(config-rpl)#done
RP/0/RP0/CPU0:XRV-5(config)#router isis 1
RP/0/RP0/CPU0:XRV-5(config-isis)#address-family ipv4 unicast
RP/0/RP0/CPU0:XRV-5(config-isis-af)#propagate level 2 into level 1
route-policy 1
```

再显示节点 1 IS-IS LSDB 的 L1 LSP 的详细信息,如下:

```
RP/0/RP0/CPU0:XRV-1#show isis d XRV-1.00-00 verbose


IS-IS 1 (Level-1) Link State Database
LSPID            LSP Seq Num   LSP Checksum   LSP Holdtime/Rcvd   ATT/P/OL
XRV-1.00-00    * 0x0000002f    0xb80f         732        /*      1/0/0
  Area Address:   01
  NLPID:          0xcc
  IP Address:     1.1.1.1
  Metric: 10         IP-Extended 1.1.1.1/32
    Prefix-SID Index: 1, Algorithm:0, R:0 N:1 P:0 E:0 V:0 L:0
    Prefix Attribute Flags: X:0 R:0 N:1
  Metric: 10         IP-Extended 12.1.1.0/28
    Prefix Attribute Flags: X:0 R:0 N:0
  Metric: 10         IP-Extended 13.1.1.0/28
    Prefix Attribute Flags: X:0 R:0 N:0
  Metric: 10         IP-Extended 17.1.1.0/28
    Prefix Attribute Flags: X:0 R:0 N:0
  Metric: 20         IP-Extended-Interarea 2.2.2.2/32
    Prefix-SID Index: 2, Algorithm:0, R:1 N:1 P:1 E:0 V:0 L:0
    Prefix Attribute Flags: X:0 R:1 N:1
  Metric: 20         IP-Extended-Interarea 3.3.3.3/32
```

```
  Prefix-SID Index: 3, Algorithm:0, R:1 N:1 P:1 E:0 V:0 L:0
  Prefix Attribute Flags: X:0 R:1 N:1
Metric: 30          IP-Extended-Interarea 4.4.4.4/32
  Prefix-SID Index: 4, Algorithm:0, R:1 N:1 P:1 E:0 V:0 L:0
  Prefix-SID Index: 1444, Algorithm:128, R:1 N:1 P:1 E:0 V:0 L:0
  Prefix-SID Index: 2444, Algorithm:129, R:1 N:1 P:1 E:0 V:0 L:0
  Prefix Attribute Flags: X:0 R:1 N:1
Metric: 40          IP-Extended-Interarea 5.5.5.5/32
  Prefix-SID Index: 5, Algorithm:0, R:1 N:1 P:1 E:0 V:0 L:0
  Prefix Attribute Flags: X:0 R:1 N:1
Metric: 30          IP-Extended-Interarea 6.6.6.6/32
  Prefix-SID Index: 6, Algorithm:0, R:1 N:1 P:1 E:0 V:0 L:0
  Prefix Attribute Flags: X:0 R:1 N:1
Metric: 40          IP-Extended-Interarea 8.8.8.8/32
  Prefix-SID Index: 8, Algorithm:0, R:1 N:1 P:1 E:0 V:0 L:0
  Prefix Attribute Flags: X:0 R:1 N:1
Metric: 50          IP-Extended-Interarea 9.9.9.9/32
  Prefix-SID Index: 9, Algorithm:0, R:1 N:1 P:1 E:0 V:0 L:0
  Prefix Attribute Flags: X:0 R:1 N:1
```

可见 L2 的路由信息，包括 SR 标签，已经汇入 L1 层路由。节点 1 的 L1 LSP 里包括 L2 网络其他节点和 L1 层区域 2 节点的 SR 标签。注意，所有 index 都包含 R-Flag，表示该 SID 是从其他 Level 或其他协议导入的。同样将 L2 index 汇入 L1 时，也会将 NP 标志位置 1。

在节点 7 上查看 MPLS 转发表，可见节点 7 已经有了到其他节点的转发项，如下：

```
RP/0/RP0/CPU0:XRV-7#show mpls forwarding
Tue Nov 23 11:07:04.298 UTC
Local  Outgoing    Prefix            Outgoing      Next Hop        Bytes
Label  Label       or ID             Interface                     Switched
------ ----------- ----------------- ------------  -----------     --------
70001  Pop         SR Pfx (idx 1)    Gi0/0/0/0     17.1.1.1        0
70002  16002       SR Pfx (idx 2)    Gi0/0/0/0     17.1.1.1        0
70003  16003       SR Pfx (idx 3)    Gi0/0/0/0     17.1.1.1        0
```

70004	16004	SR Pfx (idx 4)	Gi0/0/0/0	17.1.1.1	0
70005	16005	SR Pfx (idx 5)	Gi0/0/0/0	17.1.1.1	0
70006	16006	SR Pfx (idx 6)	Gi0/0/0/0	17.1.1.1	0
70008	16008	SR Pfx (idx 8)	Gi0/0/0/0	17.1.1.1	0
70009	16009	SR Pfx (idx 9)	Gi0/0/0/0	17.1.1.1	0

查看节点 7 路由表，可以看到节点 7 到其他节点的所有路由。如下：

```
RP/0/RP0/CPU0:XRV-7#show route ipv4

i*L1 0.0.0.0/0 [115/10] via 17.1.1.1, 07:37:36, GigabitEthernet0/0/0/0
i L1 1.1.1.1/32 [115/20] via 17.1.1.1, 01:07:30, GigabitEthernet0/0/0/0
i ia 2.2.2.2/32 [115/30] via 17.1.1.1, 01:07:30, GigabitEthernet0/0/0/0
i ia 3.3.3.3/32 [115/30] via 17.1.1.1, 01:07:30, GigabitEthernet0/0/0/0
i ia 4.4.4.4/32 [115/40] via 17.1.1.1, 01:07:30, GigabitEthernet0/0/0/0
i ia 5.5.5.5/32 [115/50] via 17.1.1.1, 01:07:30, GigabitEthernet0/0/0/0
i ia 6.6.6.6/32 [115/40] via 17.1.1.1, 01:07:30, GigabitEthernet0/0/0/0
L    7.7.7.7/32 is directly connected, 08:51:07, Loopback1
i ia 8.8.8.8/32 [115/50] via 17.1.1.1, 01:07:30, GigabitEthernet0/0/0/0
i ia 9.9.9.9/32 [115/60] via 17.1.1.1, 01:07:30, GigabitEthernet0/0/0/0
i L1 12.1.1.0/28 [115/20] via 17.1.1.1, 07:59:16, GigabitEthernet0/0/0/0
i ia 13.1.1.0/24 [115/30] via 17.1.1.1, 01:07:30, GigabitEthernet0/0/0/0
i L1 13.1.1.0/28 [115/20] via 17.1.1.1, 07:59:16, GigabitEthernet0/0/0/0
C    17.1.1.0/28 is directly connected, 08:50:21, GigabitEthernet0/0/0/0
L    17.1.1.7/32 is directly connected, 08:50:21, GigabitEthernet0/0/0/0
i ia 23.1.1.23/32 [115/30] via 17.1.1.1, 01:07:30, GigabitEthernet0/0/0/0
i ia 24.0.0.0/28 [115/30] via 17.1.1.1, 01:07:30, GigabitEthernet0/0/0/0
i ia 24.1.1.0/28 [115/30] via 17.1.1.1, 01:07:30, GigabitEthernet0/0/0/0
i ia 34.1.1.0/24 [115/30] via 17.1.1.1, 01:07:30, GigabitEthernet0/0/0/0
i ia 45.1.1.0/28 [115/40] via 17.1.1.1, 01:07:30, GigabitEthernet0/0/0/0
i ia 56.1.1.0/28 [115/40] via 17.1.1.1, 01:07:30, GigabitEthernet0/0/0/0
i ia 58.1.1.0/28 [115/50] via 17.1.1.1, 01:07:30, GigabitEthernet0/0/0/0
i ia 68.1.1.0/28 [115/40] via 17.1.1.1, 01:07:30, GigabitEthernet0/0/0/0
i ia 89.1.1.0/28 [115/50] via 17.1.1.1, 01:07:30, GigabitEthernet0/0/0/0
L    127.0.0.0/8 [0/0] via 0.0.0.0, 08:51:09
```

6.3.4 数据转发验证

在未将 L2 路由信息注入 L1 路由的情况下,L1 路由器只能知道本区域的 SR index 信息,无法知道其他区域的 SR index 信息,无法建立到其他区域节点的 MPLS 转发路径。例如,节点 7 通过路径追踪方式探测去往节点 9 的 MPLS 路径信息会失败,如下:

```
RP/0/RP0/CPU0:XRV-7#traceroute mpls ipv4 9.9.9.9/32 verbose
Q 1 * Unable to get properties from control plane
```

在将 L2 路由信息注入 L1 路由情况下,L1 路由器可以知道其他区域的 index 信息。然而,由于携带 SRGB 的 L1 LSP 不能跨区域传送,所以,一个区域内的节点无法获知其他区域节点的 PrefixSID,也就无法进行有效的路径控制和流量引导。例如,路由信息注入后,节点 7 虽然建立了去往所有节点的 MPLS 转发表项,却无法得知每一个节点的 PrefixSID 信息,这与 SR 设计理念中"PrefixSID 是局部含义,而 SID Index 是全局含义"是一致的。

在节点 7 可以通过路径追踪方式探测去往节点 9 的 MPLS 路径,信息如下:

```
RP/0/RP0/CPU0:XRV-7#traceroute mpls ipv4 9.9.9.9/32 verbose

0 17.1.1.7 17.1.1.1 MRU 1500 [Labels: 16009 Exp: 0]
L 1 17.1.1.1 13.1.1.3 MRU 1500 [Labels: 16009 Exp: 0] 109 ms, ret code 8
L 2 13.1.1.3 34.1.1.6 MRU 1500 [Labels: 16009 Exp: 0] 102 ms, ret code 8
L 3 34.1.1.6 68.1.1.8 MRU 1500 [Labels: 80009 Exp: 0] 100 ms, ret code 8
L 4 68.1.1.8 89.1.1.9 MRU 1500 [Labels: implicit-null Exp: 0] 112 ms,
ret code 8
! 5 89.1.1.9 136 ms, ret code 3
```

可以看到去往节点 9 的路径先到达 L1 区域 1 的 L1/2 路由器节点 1,依次经过 L2 区域的节点 3、节点 6,节点 6 是 L1 区域 2 的 L1/2 路由器。此时的出标签是以 L2 区域的 SRGB [16000, 23999]计算的 16009,从节点 6 开始进入 L1 区域 2,下一跳是节点 8,出标签是以节点 8 的 SRGB [80000, 81000]计算的 80009,再由节点 8 最后到达节点 9。跨区域路径正常建立成功。

受限于区域内节点只有区域拓扑视图,单纯依赖区域节点获取的信息无法实现跨区域流量调度。因此，在跨域场景中，需要引入集中控制器，控制器通过接入不同的区域获得每个区域的拓扑、节点层面的 PrefixSID 信息，甚至链路层面 AdjSID 信息。通过全局视角进行流量感知和基于 SID 的流量牵引，进而实现跨域路径编排与调度。

6.4 外部 SR 实验

通过路由重发布的方式，不同协议之间可以实现路由的双向互通。同样，通过路由重发布，也可以实现不同协议之间 PrefixSID 信息的交互。本实验主要演示将 OSPF 协议的 SID 信息传播到 IS-IS 路由域的过程。同时验证在导入 OSPF 协议的 SID 信息后，IS-IS 路由域利用 SR 标签和 OSPF 路由域通信的能力。

6.4.1 实验环境

多区域 SR 实验拓扑如图 6.12 所示，节点 1 和节点 7 之间运行 OSPF 协议，位于 OSPF 区域 0 中，节点 7 以外节点运行 IS-IS 协议，节点 2～节点 8 的配置同域间路由 SR 实验。节点 1 同时运行 OSPF 协议和 IS-IS 协议。节点 N 的 SRGB 和本地标签池配置不变。所有 OSPF 链路代价均为 1，IS-IS 链路代价均为 10，在节点 N 设置环回接口，配置 IP 地址 N.N.N.N/32，分配 SID 索引 N。节点 5 将 L2 路由信息注入 L1 区域 2。

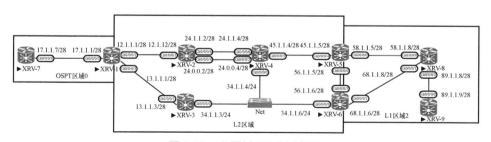

图 6.12 多区域 SR 实验拓扑

6.4.2 协议配置

节点 1 和节点 7 的配置变化如下,其他节点同域间 SR 实验。

节点 1:
```
router isis 1
 is-type level-2-only
 net 01.0000.0000.0001.00
 address-family ipv4 unicast
  metric-style wide
  segment-routing mpls
 !
 interface Loopback1
  address-family ipv4 unicast
   prefix-sid index 1
   !
  !
 interface GigabitEthernet0/0/0/2
  point-to-point
  address-family ipv4 unicast
   !
  !
 interface GigabitEthernet0/0/0/3
  point-to-point
  address-family ipv4 unicast
   !
  !
 !
router ospf 1
 router-id 1.1.1.1
 segment-routing mpls
 address-family ipv4 unicast
 area 0
```

```
  interface GigabitEthernet0/0/0/0
  !
 !
!
mpls oam
!
mpls label range table 0 100000 199999
```

节点 7:
```
router ospf 1
 router-id 7.7.7.7
 segment-routing global-block 70000 71000
 segment-routing mpls
 address-family ipv4 unicast
 area 0
  interface Loopback1
   prefix-sid index 7
  !
  interface GigabitEthernet0/0/0/0
  !
 !
!
mpls oam
!
mpls label range table 0 700000 799999
```

6.4.3 协议交互

按照上述配置，节点 1 和节点 7 建立了 OSPF 邻居。在节点 1 查看学到的 SR 标签，可以看到学到了 7.7.7.7/32 的 PrefixSID 7，如下：

```
RP/0/RP0/CPU0:XRV-1#show ospf sid-database

SID Database for ospf 1 with ID 1.1.1.1
```

```
SID            Prefix/Mask
--------       ------------------
7              7.7.7.7/32
```

在节点 1 的 IS-IS 实例中配置重发布 OSPF 路由,命令如下:

```
RP/0/RP0/CPU0:XRV-1(config)#router isis 1
RP/0/RP0/CPU0:XRV-1(config-isis)#address-family ipv4 unicast
RP/0/RP0/CPU0:XRV-1(config-isis-af)#redistribute ospf 1
```

在节点 1 显示 IS-IS LSDB 的详细信息,如下:

```
RP/0/RP0/CPU0:XRV-1#show isis d XRV-1.00-00 verbose

IS-IS 1 (Level-2) Link State Database
LSPID           LSP Seq Num   LSP Checksum  LSP Holdtime/Rcvd   ATT/P/OL
XRV-1.00-00  *  0x0000000c    0xb6e1          1077 /*            0/0/0
  Area Address:    01
  NLPID:           0xcc
  IP Address:      1.1.1.1
  Metric: 10          IP-Extended 1.1.1.1/32
    Prefix-SID Index: 1, Algorithm:0, R:0 N:1 P:0 E:0 V:0 L:0
    Prefix-SID Index: 11, Algorithm:1, R:0 N:1 P:0 E:0 V:0 L:0
    Prefix-SID Index: 1111, Algorithm:128, R:0 N:1 P:0 E:0 V:0 L:0
    Prefix-SID Index: 2111, Algorithm:129, R:0 N:1 P:0 E:0 V:0 L:0
    Prefix Attribute Flags: X:0 R:0 N:1
  Metric: 10          IP-Extended 12.1.1.0/28
    Prefix Attribute Flags: X:0 R:0 N:0
  Metric: 10          IP-Extended 13.1.1.0/28
    Prefix Attribute Flags: X:0 R:0 N:0
  Metric: 0           IP-Extended 17.1.1.0/28
    Prefix Attribute Flags: X:1 R:0 N:0
  Metric: 0           IP-Extended 7.7.7.7/32
    Prefix-SID Index: 7, Algorithm:0, R:1 N:0 P:1 E:0 V:0 L:0
    Prefix Attribute Flags: X:1 R:0 N:0
```

可以看到,节点 7 的 OSPF 的路由 7.7.7.7/32,index 7 已经自动汇入节点 1

的 L2 LSP 中。注意 index 7 包含 R-Flag，表示该 SID 是从其他 level 或其他协议导入的。

对于导入路由的 IS-IS 节点而言，从其他协议学习到的路由为非本地路由，为了避免其他节点产生虚假倒数第二跳，一定要告知与自己直连节点不能执行倒数第二跳弹出，因此，在通告中它将 NP 位置 1，而且在整个洪泛过程中，该标志都不会发生变化。

同时可见 index 7 的 N-Flag 为 0，因为它是从其他协议导入的，不是 IS-IS 节点。

所有区域内的 IS-IS 节点都将基于节点 1 的 LSP 中的信息计算得到去往 OSPF 路由 7.7.7.7/32 的转发表信息，以节点 9 为例，它到达 7.7.7.7/32 的下一跳是节点 8，路由信息如下：

```
RP/0/RP0/CPU0:XRV-9#show ip route
i ia 7.7.7.7/32 [115/50] via 89.1.1.8, 00:05:56, GigabitEthernet0/0/0/0
```

由于出标签为下一跳（节点 8）的 SRGB+SID=80000+7= 80007，而入标签则由本地 SRGB 决定，即 90000+7=90007，于是得到节点 9 的 MPLS 转发表项：

```
RP/0/RP0/CPU0:XRV-9#show mpls forwarding
Local    Outgoing    Prefix              Outgoing       Next Hop        Byte
Label    Label       or ID               Interface                      Switched
------   ---------   -----------------   ------------   ------------    --------
90007    80007       SR Pfx (idx 7)      Gi0/0/0/0      89.1.1.8        0
```

6.4.4 数据转发验证

现在已经实现了外部路由在 IS-IS 域的传播，各节点也已经建立了去往 OSPF 路由的转发表，例如，在节点 9 上通过 MPLS traceroute 可以发现单项路径已经连通：

```
RP/0/RP0/CPU0:XRV-9#traceroute mpls ipv4 7.7.7.7/32 verbose

  0 89.1.1.9 89.1.1.8 MRU 1500 [Labels: 80007 Exp: 0]
L 1 89.1.1.8 58.1.1.5 MRU 1500 [Labels: 16007 Exp: 0] 140 ms, ret code 8
L 2 58.1.1.5 56.1.1.6 MRU 1500 [Labels: 16007 Exp: 0] 143 ms, ret code 8
```

```
L 3 56.1.1.6 34.1.1.3 MRU 1500 [Labels: 16007 Exp: 0] 118 ms, ret code 8
L 4 34.1.1.3 13.1.1.1 MRU 1500 [Labels: 16007 Exp: 0] 143 ms, ret code 8
L 5 13.1.1.1 17.1.1.7 MRU 1500 [Labels: implicit-null Exp: 0] 185 ms,
ret code 8
```

可以看到去往节点 7 的路径先到达 L1 区域 2 的下一跳（节点 8），出标签是以节点 8 的 SRGB [80000, 81000] 计算的 80007，再到 L1/2 路由器节点 5，之后依次经过 L2 区域的节点 6、节点 3、节点 1。节点 1 同时是 OSPF 区域 0 的边界路由器。此时的出标签是以 L2 区域的 SRGB [16000, 23999] 计算的 16007，从节点 1 开始进入 OSPF 区域 0，下一跳是节点 7。跨协议路径建立成功。

这里可以看到节点 9 到节点 7 的路径不是最优路径。因为只有节点 5 向 L1 区域 2 注入了外部路由信息，节点 6 没有，所以节点 8 转发数据包时只能发给节点 5。如果在节点 6 里通过命令把 L2 路由信息注入 L1 区域 2，如下：

```
RP/0/RP0/CPU0:XRV-6(config)#route-policy 1
RP/0/RP0/CPU0:XRV-6(config-rpl)# done
RP/0/RP0/CPU0:XRV-6(config)#router isis 1
RP/0/RP0/CPU0:XRV-6(config-isis)#address-family ipv4 unicast
RP/0/RP0/CPU0:XRV-6(config-isis-af)#propagate level 2 into level 1
route-policy 1
```

此时继续在节点 9 上通过 MPLS traceroute 可以发现路径如下：

```
RP/0/RP0/CPU0:XRV-9#traceroute mpls ipv4 7.7.7.7/32 verbose

0 89.1.1.9 89.1.1.8 MRU 1500 [Labels: 80007 Exp: 0]
L 1 89.1.1.8 68.1.1.6 MRU 1500 [Labels: 16007 Exp: 0] 81 ms, ret code 8
L 2 68.1.1.6 34.1.1.3 MRU 1500 [Labels: 16007 Exp: 0] 65 ms, ret code 8
L 3 34.1.1.3 13.1.1.1 MRU 1500 [Labels: 16007 Exp: 0] 58 ms, ret code 8
L 4 13.1.1.1 17.1.1.7 MRU 1500 [Labels: implicit-null Exp: 0] 72 ms,
ret code 8
```

可见节点 8 直接把数据包转发给了节点 6，没有再绕过节点 5，优化了路径选择。这里体现了当 IS-IS 区域存在多个区域边界 L1/2 路由器时，配置 L1/2 路由器把 L2 路由信息注入 L1 路由区域，可以优化 L1 路由器的出口选择。

第 7 章

SR BGP 原理与实战

7.1 SR BGP 基本原理

7.1.1 基本概念

BGP 可作为 SR 的控制面，用来进行 SID 的分配和分发。本章涉及了两类与 BGP 相关的 SID：BGP PrefixSID 和 BGP PeeringSID。其中，BGP PrefixSID 与某个 IP 前缀相关联，在网络中全局有效。BGP PeeringSID 与 BGP 对等体会话的特定邻居关联，具体可细分为 PeerNodeSID、PeerAdjSID、PeerSetSID 3 种，PeeringSID 通过 BGP 的出口对等体工程（Egress Peer Engineering，EPE）功能进行自动分配，通常只在某个节点上局部有效。

本章阐述两种 SID 分发和使用方式。节点分配的 PrefixSID 信息通过 BGP 标签单播（BGP Labeled Unicast，BGP-LU）传递，PeeringSID 通过 BGP 链路状态（BGP Link State，BGP-LS）进行通告。

7.1.2 基于 BGP-LU 的 PrefixSID 通告机制

BGP-LU 既可以用于外部边界网关协议（External Border Gateway Protocol，EBGP）也可用于内部边界网关协议（Internal Border Gateway Protocol，IBGP）。

BGP-LU 的实现细节由 IETF RFC3107 定义，采用随机分配标签的方式，为网络前缀分配标签，并与 LDP 配合共同实现对 BGP/MPLS L3VPN 功能的支持。BGP-LU 在 BGP 原有单播/多播等子地址族基础上拓展了 Labeled Unicast 地址族（SAFI=4），相关信息放置在 MP_REACH_NLRI 属性的网络可达信息（Network Layer Reachability Information，NLRI）中。Labeled Unicast NLRI 采用一个三元组结构，包含了长度、一个或者多个标签以及前缀信息。BGP IPv4 Labeled Unicast NLRI 如图 7.1 所示。

图 7.1　BGP IPv4 Labeled Unicast NLRI

BGP-LU 的随机标签分配机制，难以满足 SR 机制对 SID 的控制以及域内域间 SID 配合协作需求，因此，在 RFC8669 中，SR 拓展了 BGP-LU 以通告 BGP PrefixSID，定义了 PrefixSID 属性（属性类型 40）和 SRGB 属性（属性类型 41），前者承载 PrefixSID 信息，后者承载通告节点的 SRGB 信息。

PrefixSID 携带了一个 Label-Index 结构，BGP-LS PrefixSID 属性如图 7.2 所示，其中 Type 域的值固定为 1，RESERVED 域和 Flags 域目前没有实际用途，固定为 0。Label-Index 是长度为 4byte 的标签索引。

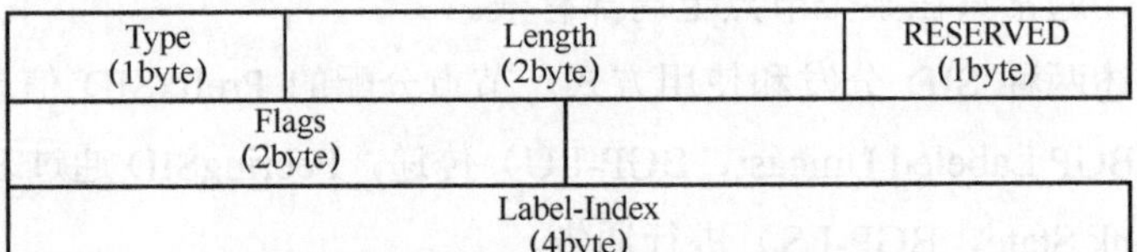

图 7.2　BGP-LS PrefixSID 属性

在启动 SR-BGP 之前，本地为 BGP Prefix 分配的标签由本地随机生成。在启动 SR-BGP 后，对于源发前缀的节点而言，本地标签由管理员手工指定；对于其他节点，则基于 SRGB 和从邻居收到的 SID 索引计算得到。

7.1.3 基于 BGP-LS 的 PeeringSID 通告机制

BGP PeeringSID 经常用于集中式流量工程。集中式流量工程采用中心计算方式，由路径计算单元（Path Computation Element，PCE）收集全网节点 SID、链路 SID 等信息，从全局角度掌握网络拓扑，按照预先设定的流量路由策略，计算编排路径形成 SID 列表，在流量进入头节点时压入 SID 列表，其他节点按照 SID 列表引导流量经过特定节点，使得流量按照要求经过预期节点到达目的地。与 IGP 为邻居链路分配 AdjSID 一样，BGP 通过 EPE 机制为邻居节点或链路分配 PeeringSID。

RFC8402 定义了 3 种类型的 BGP PeeringSID：PeerNode SID、PeerAdj SID、PeerSet SID。

- Peer Node Segment：描述了一个对等（Peer）节点的 Segment。
- Peer Adjacency Segment：描述了一条链路的 Segment。
- Peer Set Segment：描述了一个链路和节点的集合的 Segment。

BGP 拓展了 NLRI 属性并新定义了 BGP-LS 属性，前者用于描述链路状态，后者用于描述对应的链路标签信息。

1. 拓展的 BGP-LS NLRI

RFC7752 定义的 BGP-LS NLRI 信息格式，其中 Link NLRI（NLRI Type = 2）格式如图 7.3 所示，包括本地节点、远端节点以及链路描述符信息。BGP 节点描述符分为本地节点描述符 TLV（Type=256）和远端节点描述符（Type=257）。这两种节点描述格式一样，必须包含节点的 BGP Router 标识符（TLV 516）、自治系统（Autonomous System，AS）号码（TLV 512）信息；链路描述符包含本地链路描述符和邻居链路描述符两个字段，这两个字段可以采用 IP 地址或者链路 ID 的描述方式，前者采用 IP 地址、后者采用链路 ID 来描述链路端点信息，详细定义参考 RFC7752。

Protocol-ID (1byte)
Identifier (8byte)
Local Node Descriptors (4byte)
Remote Node Descriptors (4byte)
Link Descriptors (4byte)

图 7.3 BGP-LS Link NLRI 格式

2. BGP-LS 属性

RFC9086 定义了 BGP-LS 属性（属性类型 29），该属性的格式具体定义如图 7.4 所示。

Type (2byte)		Length (2byte)
Flags (1byte)	Weight (1byte)	Reserved (2byte)
SID/Label/Index (可变长度)		

图 7.4　BGP-LS 属性格式具体定义

Type 域可以为 1101、1102 和 1103，分别代表 PeerNode SID、PeerAdj SID 和 PeerSet SID。

Flags 字段定义了 4 个标志位：V=1 表示携带 SID 标签值，L=1 表示使用本地标签；B=1 用于指示 FRR 备用路径；P=1 表示标签持久化，即使该路由器发生了重启也会改变为该链路分配的 SID。默认情况下 V 和 L 均设置为 1。

SID/Label/Index 域以 TLV 格式记录为 PeerNode、PeerAdj 或 PeerSet 分配的 SID 信息。

- 如果是 3byte，则表示标签值，最右端 20bit 代表了标签值，此时 V 和 L 标志必须置 1。
- 如果是 4byte，则表示 SID index 值，此时要一并通告 SRGB 信息。

7.2　基于 BGP-LU 的标签分发

在 BGP-SR 之前，BGP 已经具备为前缀分配标签的能力。为了支持 BGP/MPLS L3VPN 机制，BGP 拓展了 Labeled Unicast 地址族，通过 BGP-LU 在相互之间交互 BGP 前缀标签。先通过这个实验来理解基本的 BGP 分发机制、BGP 产生的 MPLS 交换表项与 IGP 路由的关联关系，为更好地理解 BGP-SR 机制奠定基础。

7.2.1　实验环境

BGP-LU 标签分配实验拓扑如图 7.5 所示，所有节点均已按第 2 章的实验约定配置了接口 IP 地址、环回接口地址等。实验环境共包括 7 台路由器，其中节点 2～节点 5 属于 AS10，域内运行 OSPF 协议，并启动 LDP 分发域内 MPLS 标签，这些节点的 Loopback 地址在 OSPF 域内发布；节点 2、节点 4、节点 5 运行全互联 IBGP。

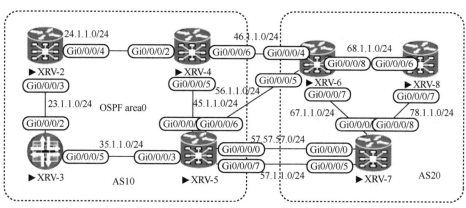

图 7.5　BGP-LU 标签分配实验拓扑

节点 6～节点 8 组成 AS20，因为它们相互之间通过 BGP 直连，域内暂时不运行 IGP；节点 6～节点 8 运行全互联 IBGP；节点 4、节点 6 运行 EBGP，节点 5 与节点 6、节点 7 运行 EBGP，均采用直连接口地址建立 EBGP 会话。

节点 N 的本地标签空间为 $N00000$～$N99999$。

7.2.2　基本配置

节点 4 的关键配置如下：
```
route-policy bgp_in
  pass
end-policy
!
route-policy bgp_out
  pass
end-policy
!
```

```
router ospf 1
 ...//OSPF 协议配置省略
!
router bgp 10
 address-family ipv4 unicast
  allocate-label all
 !
 neighbor 2.2.2.2
  remote-as 10
  update-source Loopback0
  address-family ipv4 labeled-unicast
   route-policy bgp_in in
   route-policy bgp_out out
   next-hop-self
  !
 !
 neighbor 5.5.5.5
  remote-as 10
  update-source Loopback0
  address-family ipv4 labeled-unicast
   route-policy bgp_in in
   route-policy bgp_out out
   next-hop-self
  !
 !
 neighbor 46.1.1.6
  remote-as 20
  address-family ipv4 labeled-unicast
   route-policy bgp_in in
   route-policy bgp_out out
  !
 !
!
```

```
mpls oam
!
mpls ldp
 interface GigabitEthernet0/0/0/2
 !
 interface GigabitEthernet0/0/0/5
 !
!
mpls label range table 0 400000 499999
```

节点 6 的关键配置如下：
```
route-policy bgp_in
  pass
end-policy
!
route-policy bgp_out
  pass
end-policy
!
router bgp 20
 address-family ipv4 unicast
  network 6.6.6.6/32
  allocate-label all
 !
 neighbor 46.1.1.4
  remote-as 10
  address-family ipv4 labeled-unicast
   route-policy bgp_in in
   route-policy bgp_out out
  !
 !
 neighbor 56.1.1.5
  remote-as 10
  address-family ipv4 labeled-unicast
   route-policy bgp_in in
```

```
   route-policy bgp_out out
  !
 !
 neighbor 67.1.1.7
  remote-as 20
  address-family ipv4 labeled-unicast
   route-policy bgp_in in
   route-policy bgp_out out
   next-hop-self
  !
 !
 neighbor 68.1.1.8
  remote-as 20
  address-family ipv4 labeled-unicast
   route-policy bgp_in in
   route-policy bgp_out out
   next-hop-self
  !
 !
!
mpls oam
!
mpls label range table 0 600000 699999
end
```

其他节点的 OSPF 配置和 LDP 配置可参阅节点 4 和节点 6。需要注意的是 AS10 内节点之间的 IBGP 会话采用 Loopback 接口地址建立，AS20 内节点之间的 IBGP 会话采用物理接口地址建立。

7.2.3 BGP-LU 机制

1. 使用 Loopback 地址建立 IBGP 会话

首先，由于 AS10 运行 OSPF 协议，节点之间拥有 Loopback 地址路由，因此，节点 2、节点 4、节点 5 拥有彼此 Loopback 地址的路由，相互之间可以通过 Loopback 接口地址建立 IBGP 会话，这也是 IBGP 通常的使用方式。使用 Loopback 地址建立 IBGP 会话可以使得 BGP 会话状态不依赖于接口状态。例如，如果节点 4 的 Gi0/0/0/2 接口关闭，但是还有一条经过节点 3 的路由使得它到达节点 2 的 Loopback 地址，这样 BGP 会话就会一直保持 Established 状态。

BGP 采用 neighbor 指令设定发起 BGP 会话的目的 IP 地址，这个地址被设定为 IBGP 邻居的 Loopback 接口地址，由于会话的双向性，这个 BGP 会话的源地址最好使用本地的 Loopback 接口地址，这样就可以提高会话的鲁棒性。BGP 会话的默认源地址是基于本地路由自动填写的。例如，当节点 5 准备与节点 2 的 Loopback 地址建立 IBGP 会话时，会使用目的 IP 地址匹配本地路由表，发现要通过本地接口 Gi0/0/0/2 去往 2.2.2.2，那么 BGP 会话的源地址就会填写该接口地址 24.1.1.4。这自然不是管理员期望的，为了改变这种默认的源 IP 地址选用方式，需要以下指令：

```
router bgp <asn> neighbor <a.b.c.d> update-source Loopback0
```

对于 AS20 内节点而言，彼此之间都是全互联，可以通过直连路由互通，因此，暂时没有启动 IGP。由于没有 IGP 发布 Loopback 地址，这些节点之间使用直连接口地址建立 IBGP 会话。

2．IBGP 路由通告时下一跳地址和标签的修正

按照 BGP 规定，IBGP 在进行路由通告时不修改 BGP 的 Nexthop 属性字段，即不改变下一跳地址。当节点 4 和节点 5 从 EBGP 邻居收到外部路由 8.8.8.8/32 后，将该路由通告给节点 2，查看节点 2 收到的路由和标签信息，如下：

```
RP/0/0/CPU0:XRV-2#show bgp 8.8.8.8/32
BGP routing table entry for 8.8.8.8/32
Versions:
  Process             bRIB/RIB     SendTblVer
  Speaker                   28             28
    Local Label: 200007
Paths: (2 available, no best path)
  Not advertised to any peer
  Path #1: Received by speaker 0
  Not advertised to any peer
```

```
     20
       46.1.1.6 (inaccessible) from 4.4.4.4 (4.4.4.4)
         Received Label 600000
         Origin IGP, localpref 100, valid, internal, labeled-unicast
         Received Path ID 0, Local Path ID 0, version 0
     Path #2: Received by speaker 0
     Not advertised to any peer
     20
       56.1.1.6 (inaccessible) from 5.5.5.5 (5.5.5.5)
         Received Label 600000
         Origin IGP, localpref 100, valid, internal, labeled-unicast
         Received Path ID 0, Local Path ID 0, version 0
RP/0/0/CPU0:XRV-2#show ip route 8.8.8.8/32
% Network not in table
```

可以看到节点 2 收到节点 4(4.4.4.4)和节点 5(5.5.5.5)通告的关于 8.8.8.8/32 的路由信息，其下一跳为节点 6 的不同 EBGP 会话地址，标签信息也是节点 6 为该前缀分配的标签，而不是节点4或节点5为该路由分配的标签，这是为了保证下一跳与标签的一致性。由于 AS10 内没有到该地址的路由，因此，这条 BGP 路由被标记为"inaccessible"。可以通过路由重发布或者静态路由等方式让节点 2 建立到达 46.1.1.6 或者 56.1.1.6 的路由（即使建立了路由，由于标签跟下一跳不匹配，也会导致 MPLS 转发不生效）。但为了保持域内路由的稳定性，通常不会这样做。通常的做法是在向特定邻居通告路由时，强制 BGP 进程修改 Nexthop 属性字段，用到的指令为：

```
router bgp <asn> neighbor <a.b.c.d> address-family ipv4 labeled-
unicast next-hop-self
```

在所有IBGP 会话上配置上述指令后，再次查看节点 2 收到的关于外部路由 8.8.8.8/32 的路由信息，如下：

```
RP/0/0/CPU0:XRV-2#show bgp 8.8.8.8/32
BGP routing table entry for 8.8.8.8/32
Versions:
  Process              bRIB/RIB    SendTblVer
```

```
Speaker                   37            37
   Local Label: 200007
Paths: (2 available, best #1)
  Not advertised to any peer
  Path #1: Received by speaker 0
  Not advertised to any peer
  20
    4.4.4.4 (metric 2) from 4.4.4.4 (4.4.4.4)
      Received Label 400008
      Origin IGP, localpref 100, valid, internal, best, group-
best, labeled-unicast
      Received Path ID 0, Local Path ID 0, version 37
  Path #2: Received by speaker 0
  Not advertised to any peer
  20
    5.5.5.5 (metric 3) from 5.5.5.5 (5.5.5.5)
      Received Label 500011
      Origin IGP, localpref 100, valid, internal, labeled-unicast
      Received Path ID 0, Local Path ID 0, version 0

RP/0/0/CPU0:XRV-2#show ip route 8.8.8.8/32
Routing entry for 8.8.8.8/32
  Known via "bgp 10", distance 200, metric 0, [ei]-bgp
  Tag 20, type internal
  Routing Descriptor Blocks
    4.4.4.4, from 4.4.4.4
      Route metric is 0
  No advertising protos.
```

可以看到，节点 4 和节点 5 在通告 8.8.8.8/32 路由时分别将下一跳地址修改为对应的 IBGP 会话地址，标签信息为通告节点为该前缀分配的本地标签。

3. BGP-LU 的标签分配

与 IGP 自动发现路由不同，BGP 默认不通告任何路由信息。BGP 通告的路由由管理员明确配置。管理员可以通过重发布方式，将 IGP 发现的路由宣告到

BGP 中，也可以明确告知 BGP 宣告哪些路由。宣告的路由要在对应地址族中配置，指令为：

```
router bgp <asn> address-family ipv4 unicast network <a.b.c.d/M>
```

为 BGP 路由分配标签的指令为：

```
router bgp <asn> address-family ipv4 unicast allocate-label [all |
route-policy WORD]
```

实际应用中往往不需要为本地发布的所有 BGP 路由分配标签，此时需要使用 route-policy 进行路由过滤。为了简化实验，使用 "allocate-label all" 指令让 BGP 为所有路由分配标签。

明确了为哪些路由分配标签后，需要设定向哪些 BGP 邻居通告这些标签信息。BGP-LU 的配置模式如下：

```
route-policy bgp_in
  pass
end-policy
!
route-policy bgp_out
  pass
end-policy
!
router bgp <asn>
  neighbor <a.b.c.d>
  address-family ipv4 labeled-unicast
  route-policy bgp_in in
  router-policy bgp_out out
```

通常情况下 BGP 发布的路由规模很大，而往往只有 L3VPN 相关路由才需要为之分配标签。因此在交互标签信息时，除了要明确指定与哪些邻居建立 BGP-LU 会话，还需要进一步指定愿意接受邻居发过来的哪些路由的标签信息，以及向邻居通告哪些路由的标签信息。通过 route-policy 策略可精准地将这些路由过滤。由于这不是本文重点，所以，实验中不对引入和通告的标签做策略控制。默认情况下，邻居之间建立了 BGP-LU 会话后不会自动通告标签信息，必须指定路由策略，哪怕路由策略内容为空，这就是实验中配置空的路由策略的原因。

要注意 IPv4 Unicast 与 IPv4 labeled Unicast 地址族的不同。当在邻居之间启动 IPv4 Unicast 地址族后，只会发布路由信息，而不会发布标签信息。启动 IPv4 Labeled Unicast 地址族，则同时发布 BGP 路由和标签信息。也就是说，虽然 BGP-LU 仅用于分配标签信息，但同时具备路由发布的功能，因为标签需要绑定到路由上才有意义。前文已阐述过，BGP-LU 是拓展了 BGP 的 MP_REACH_NLRI 属性，一并通告路由和标签信息。BGP-LU 报文示例如图 7.6 所示，节点 4 在发布 4.4.4.4/32 的标签信息时，通过 BGP 的 MP_REACH_NLRI 来承载，AFI 为 IPv4，SAFI 为 Labeled Unicast，同时携带前缀信息和标签信息。因此，启动 IPv4 Labeled Unicast 地址族分发标签时，不需要再启动 IPv4 Unicast 地址族来发布路由。

```
▶ Internet Protocol Version 4, Src: 4.4.4.4, Dst: 2.2.2.2
▶ Transmission Control Protocol, Src Port: 179, Dst Port: 53330, Seq: 134, Ack: 77, Len: 65
▼ Border Gateway Protocol - UPDATE Message
    Marker: ffffffffffffffffffffffffffffffff
    Length: 65
    Type: UPDATE Message (2)
    Withdrawn Routes Length: 0
    Total Path Attribute Length: 42
  ▼ Path attributes
    ▼ Path Attribute - MP_REACH_NLRI
      ▶ Flags: 0x90, Optional, Extended-Length, Non-transitive, Complete
        Type Code: MP_REACH_NLRI (14)
        Length: 17
        Address family identifier (AFI): IPv4 (1)
        Subsequent address family identifier (SAFI): Labeled Unicast (4)
      ▼ Next hop network address (4 bytes)
          Next Hop: 4.4.4.4
        Number of Subnetwork points of attachment (SNPA): 0
      ▼ Network layer reachability information (8 bytes)
        ▼ Label Stack=3 (bottom) IPv4=4.4.4.4/32
            MP Reach NLRI Prefix length: 56
            MP Reach NLRI Label Stack: 3 (bottom)
            MP Reach NLRI IPv4 prefix: 4.4.4.4
    ▼ Path Attribute - ORIGIN: IGP
      ▶ Flags: 0x40, Transitive, Well-known, Complete
        Type Code: ORIGIN (1)
        Length: 1
        Origin: IGP (0)
    ▼ Path Attribute - AS_PATH: empty
      ▶ Flags: 0x40, Transitive, Well-known, Complete
        Type Code: AS_PATH (2)
        Length: 0
    ▼ Path Attribute - MULTI_EXIT_DISC: 0
      ▶ Flags: 0x80, Optional, Non-transitive, Complete
        Type Code: MULTI_EXIT_DISC (4)
        Length: 4
        Multiple exit discriminator: 0
    ▼ Path Attribute - LOCAL_PREF: 100
      ▶ Flags: 0x40, Transitive, Well-known, Complete
        Type Code: LOCAL_PREF (5)
        Length: 4
        Local preference: 100
```

图 7.6　BGP-LU 报文示例

4. 基于 EBGP 的 MPLS 转发表项生效机制

EBGP 会话对等体之间通常是直连的，然而，BGP-LU 不会迭代直连路由来使得路由和 MPLS 转发表项生效。

节点 4 从 EBGP 邻居节点 6 和 IBGP 邻居节点 5 收到关于 8.8.8.8/32 的路由标签信息，如下：

```
RP/0/0/CPU0:XRV-4#show bgp 8.8.8.8/32
BGP routing table entry for 8.8.8.8/32
Versions:
  Process           bRIB/RIB     SendTblVer
  Speaker              21            21
    Local Label: 400008
Paths: (2 available, best #2)
  Advertised to update-groups (with more than one peer):
    0.3
  Path #1: Received by speaker 0
  Not advertised to any peer
  20
    5.5.5.5 (metric 2) from 5.5.5.5 (5.5.5.5)
      Received Label 500011
      Origin IGP, localpref 100, valid, internal, labeled-unicast
      Received Path ID 0, Local Path ID 0, version 0
  Path #2: Received by speaker 0
  Advertised to update-groups (with more than one peer):
    0.3
  20
    46.1.1.6 from 46.1.1.6 (6.6.6.6)
      Received Label 600000
      Origin IGP, localpref 100, valid, external, best, group-best, labeled-unicast
      Received Path ID 0, Local Path ID 0, version 21
      Origin-AS validity: not-found
--------------------------------------------------------------------
RP/0/0/CPU0:XRV-4#show ip route 8.8.8.8/32
```

```
Routing entry for 8.8.8.8/32
  Known via "bgp 10", distance 20, metric 0, [ei]-bgp, labeled unicast
(3107)
  Tag 20, type external
  Routing Descriptor Blocks
    46.1.1.6, from 46.1.1.6, BGP external
      Route metric is 0
  No advertising protos.
-----------------------------------------------------------------
RP/0/0/CPU0:XRV-4#show mpls forwarding
Sat Oct 29 11:03:58.122 UTC
Local    Outgoing   Prefix       Outgoing       Next Hop      Byte
Label    Label      or ID        Interface                    Switched
------   --------   ----------   ------------   -----------   --------
400001   Pop        2.2.2.2/32   Gi0/0/0/2      24.1.1.2      74609
400002   Pop        23.1.1.0/24  Gi0/0/0/2      24.1.1.2      520
400003   200002     3.3.3.3/32   Gi0/0/0/2      24.1.1.2      0
         500005     3.3.3.3/32   Gi0/0/0/5      45.1.1.5      0
400004   Pop        35.1.1.0/24  Gi0/0/0/5      45.1.1.5      400
400005   Pop        5.5.5.5/32   Gi0/0/0/5      45.1.1.5      46613
400006   Pop        6.6.6.6/32                  46.1.1.6      0
400007   600001     7.7.7.7/32                  46.1.1.6      0
400008   600000     8.8.8.8/32                  46.1.1.6      189880
-----------------------------------------------------------------
RP/0/0/CPU0:XRV-4#ping mpls ipv4 8.8.8.8/32 source 4.4.4.4
Sending 5, 100-byte MPLS Echos to 8.8.8.8/32,
    timeout is 2 seconds, send interval is 0 msec:

Codes: '!' - success, 'Q' - request not sent, '.' - timeout,
  'L' - labeled output interface, 'B' - unlabeled output interface,
  'D' - DS Map mismatch, 'F' - no FEC mapping, 'f' - FEC mismatch,
  'M' - malformed request, 'm' - unsupported tlvs, 'N' - no rx label,
  'P' - no rx intf label prot, 'p' - premature termination of LSP,
  'R' - transit router, 'I' - unknown upstream index,
```

```
'X' - unknown return code, 'x' - return code 0
Type escape sequence to abort.

QQQQQ
Success rate is 0 percent (0/5)
```

从上述信息可以看到，按照 BGP 选路策略，与 IBGP 路由相比，BGP 优选 EBGP。因此，节点 4 去往 8.8.8.8/32 的优选下一跳为节点 6（46.1.1.6），并选用节点 6 分发的标签 600000 作为出标签。然而，从 MPLS 转发表项看到，该表项的出接口为空。这意味着需要进一步进行路由迭代以得到出接口信息。此时发送 MPLS Echo 报文，出现"request not sent"错误，正是因为转发表项信息不完善，导致本地无法发送 MPLS Echo 报文。虽然节点 4 到 46.1.1.6 是直连网络，但 BGP-LU 却无法自动迭代得到出接口，这说明 BGP-LU 是不会迭代直连路由来计算出接口的。为了解决这个问题，要配置指向下一跳 IP 地址的静态路由，以触发 IOS 为之生成 MPLS 转发表项，供 EBGP MPLS 转发表项迭代。静态路由配置指令如下：

```
router static address-family ipv4 unicast <a.b.c.d/Mask> out_interface
```

因为目的地是下一跳节点 IP 地址，因此，配置这条静态路由时不需要指定下一跳 IP 地址，IOS 会因为没有设置下一跳而告警，但不会影响路由的生效。IOS 会依据配置的静态路由和 EBGP 会话信息自动生成特殊的 MPLS 转发表项，供 BGP-LU 迭代使用。在节点 4 上配置如下静态路由，并进一步查看 MPLS 转发表项如下：

```
RP/0/0/CPU0:XRV-4(config)#router static address-family ipv4
unicast 46.1.1.6/32 Gi0/0/0/6
RP/0/0/CPU0:XRV-4(config)#commit
RP/0/0/CPU0:Oct 29 11:21:19.611 : ipv4_static[1041]: %ROUTING-
IP_STATIC-4-CONFIG_NEXTHOP_ETHER_INTERFACE : Route for 46.1.1.6 is
configured via ethernet interface without nexthop, Please check if
this is intended
--------------------------------------------------------------------------
RP/0/0/CPU0:XRV-4#show mpls forwarding
Sat Oct 29 11:12:23.867 UTC
```

```
Local     Outgoing    Prefix        Outgoing       Next Hop      Bytes
Label     Label       or ID         Interface                    Switched
------    --------    ----------    -----------    ----------    --------
400000    Pop         46.1.1.6/32   Gi0/0/0/6      46.1.1.6      0
400001    Pop         2.2.2.2/32    Gi0/0/0/2      24.1.1.2      77174
400002    Pop         23.1.1.0/24   Gi0/0/0/2      24.1.1.2      520
400003    200002      3.3.3.3/32    Gi0/0/0/2      24.1.1.2      0
          500005      3.3.3.3/32    Gi0/0/0/5      45.1.1.5      0
400004    Pop         35.1.1.0/24   Gi0/0/0/5      45.1.1.5      400
400005    Pop         5.5.5.5/32    Gi0/0/0/5      45.1.1.5      48425
400006    Pop         6.6.6.6/32    Gi0/0/0/6      46.1.1.6      0
400007    600001      7.7.7.7/32    Gi0/0/0/6      46.1.1.6      0
400008    600000      8.8.8.8/32    Gi0/0/0/6      46.1.1.6      187860
```

可以看到，在配置了静态路由后，IOS 自动生成了一条关于 46.1.1.6/32 的 MPLS 转发表项，其他的 EBGP 路由通过迭代这条转发表项，生成了拥有出接口信息的有效 MPLS 转发表项。

在 AS10 和 AS20 之间的 EBGP 路由都需要增加相应的静态路由配置，以使能来自 EBGP 的 MPLS 转发表项。

5. 基于 IBGP 的 MPLS 转发表项生效机制

对于 IBGP 会话而言，两个 IBGP 邻居不一定直接相连，但可以借助域内 IGP 实现两个节点间互通，而且 LDP 也为所有节点生成了有效的 MPLS 转发表项。虽然 BGP-LU 生成 MPLS 转发表项时出接口为空，但是可以通过迭代已经存在的 MPLS 转发表项而得到相应的出接口，进而使得 MPLS 转发表项有效。查看节点 2 的 MPLS 转发表项信息：

```
RP/0/0/CPU0:XRV-2#show mpls forwarding
Local     Outgoing    Prefix        Outgoing       Next Hop      Bytes
Label     Label       or ID         Interface                    Switched
------    --------    ----------    -----------    ----------    --------
200000    Pop         4.4.4.4/32    Gi0/0/0/4      24.1.1.4      53461
200001    Pop         45.1.1.0/24   Gi0/0/0/4      24.1.1.4      0
200002    Pop         3.3.3.3/32    Gi0/0/0/3      23.1.1.3      263
200003    Pop         35.1.1.0/24   Gi0/0/0/3      23.1.1.3      0
```

```
200004    300000    5.5.5.5/32    Gi0/0/0/3    23.1.1.3    26780
          400005    5.5.5.5/32    Gi0/0/0/4    24.1.1.4    0
200005    400006    6.6.6.6/32                 4.4.4.4     0
200006    400007    7.7.7.7/32                 4.4.4.4     0
200007    400008    8.8.8.8/32                 4.4.4.4     0
------------------------------------------------------------------
RP/0/0/CPU0:XRV-2#ping mpls ipv4 8.8.8.8/32 source 2.2.2.2

Sending 5, 100-byte MPLS Echos to 8.8.8.8/32,
     timeout is 2 seconds, send interval is 0 msec:

Codes: '!' - success, 'Q' - request not sent, '.' - timeout,
  'L' - labeled output interface, 'B' - unlabeled output interface,
  'D' - DS Map mismatch, 'F' - no FEC mapping, 'f' - FEC mismatch,
  'M' - malformed request, 'm' - unsupported tlvs, 'N' - no rx label,
  'P' - no rx intf label prot, 'p' - premature termination of LSP,
  'R' - transit router, 'I' - unknown upstream index,
  'X' - unknown return code, 'x' - return code 0

Type escape sequence to abort.

NNNNN
Success rate is 0 percent (0/5)
```

可以看到，节点 2 优选节点 4 通告的外部路由信息，同时，本地已经通过 LDP 生成了到达 4.4.4.4/32 的有效 MPLS 转发表项。依托路由的迭代查找可以使这些外部路由转发表项有效。从节点 2 能够正常发出 MPLS Echo 报文就可以证明这一点（这里节点没有得到正确的回应是因为去往节点 8 的中间节点存在异常 MPLS 转发表项，这个问题会在后文得到解决）。

位于 AS20 内的节点并未运行 IGP，查看节点 8 的 MPLS 转发表项信息：

```
RP/0/0/CPU0:XRV-8#show mpls forwarding
Local    Outgoing    Prefix           Outgoing       Next Hop       Bytes
Label    Label       or ID            Interface                     Switched
------   --------    --------------   -----------    ----------    --------
```

800000	Pop	6.6.6.6/32		8.1.1.6	0
800001	Pop	7.7.7.7/32		78.1.1.7	0
800002	600004	2.2.2.2/32		68.1.1.6	0
800003	600005	4.4.4.4/32		68.1.1.6	0
800004	600006	5.5.5.5/32		68.1.1.6	0

可以看到，BGP-LU 生成的 MPLS 转发表项出接口依然为空，这说明 BGP-LU 不迭代直连路由来生成出接口信息。那么 BGP-LU 是否可以借鉴 EBGP 的 MPLS 转发表项生效方式，通过配置静态路由方式来使能 MPLS 转发表项呢？

```
RP/0/0/CPU0:XRV-8(config)#router static address-family ipv4 u 68.1.1.6/32 g0/0/0/6
RP/0/0/CPU0:XRV-8(config)#do show mpls forwarding
```

Local Label	Outgoing Label	Prefix or ID	Outgoing Interface	Next Hop	Bytes Switched
------	------------	-------------	------------	------------	--------
800000	Pop	6.6.6.6/32	Gi0/0/0/6	68.1.1.6	3822
800001	Pop	7.7.7.7/32	Gi0/0/0/7	78.1.1.7	5976
800002	600004	2.2.2.2/32		68.1.1.6	0
800003	600005	4.4.4.4/32		68.1.1.6	0
800004	600006	5.5.5.5/32		68.1.1.6	0
800006	Unlabelled	68.1.1.6/32	Gi0/0/0/6	point2point	0

通过配置静态路由，IOS 自动生成了一条 MPLS 转发表项，但是这条转发表项的操作却是"Unlabelled"，这说明本地认为下一跳节点不支持 MPLS，因此，静态路由无法使 IBGP 邻居通告的外部 MPLS 转发表项有效。这意味着 IBGP 通告的 MPLS 转发表项只能通过迭代 IGP 路由才能生效。因此，需要改变 AS20 内节点的配置，在其内部运行 IGP。

采用与 AS10 类似的配置方式，在节点 6～节点 8 之间运行 OSPF 协议和 LDP，再查看节点 8 的 MPLS 转发表项信息：

```
RP/0/0/CPU0:XRV-8#show mpls forwarding
```

Local Label	Outgoing Label	Prefix or ID	Outgoing Interface	Next Hop	Bytes Switched
------	------------	-------------	------------	------------	--------

```
800000       Pop          6.6.6.6/32                         68.1.1.6       0
800001       Pop          7.7.7.7/32                         78.1.1.7       0
800002       600004       2.2.2.2/32                         68.1.1.6       0
800003       600005       4.4.4.4/32                         68.1.1.6       0
800004       600006       5.5.5.5/32                         68.1.1.6       0
800005       Pop          67.1.1.0/24       Gi0/0/0/6        68.1.1.6       0
             Pop          67.1.1.0/24       Gi0/0/0/7        78.1.1.7       0
```

可以看到，由于 OSPF 生成的路由中并不包含 68.1.1.6/32 前缀，因此，LDP 也无法为该前缀生成标签信息。看来简单运行 IGP 也无法为外部路由标签的迭代生效提供支撑。

为了解决这个问题，必须重新调整 AS20 内 IBGP 会话建立方式，采用与 AS10 相似的配置模式，即采用 Loopback 地址建立 IBGP 会话。重新配置节点 6～节点 8 的 IBGP 会话后，查看节点 8 的 MPLS 转发表项信息：

```
RP/0/0/CPU0:XRV-8#show mpls forwarding
Local    Outgoing     Prefix            Outgoing         Next Hop       Bytes
Label    Label        or ID             Interface                       Switched
------   ----------   ---------------   ------------     -----------    --------
800000   Pop          6.6.6.6/32        Gi0/0/0/6        68.1.1.6       8747
800001   Pop          7.7.7.7/32        Gi0/0/0/7        78.1.1.7       10994
800002   600004       2.2.2.2/32                         6.6.6.6        0
800003   600005       4.4.4.4/32                         6.6.6.6        0
800004   600006       5.5.5.5/32                         6.6.6.6        0
800005   Pop          67.1.1.0/24       Gi0/0/0/6        68.1.1.6       0
         Pop          67.1.1.0/24       Gi0/0/0/7        78.1.1.7       0
------------------------------------------------------------------------------
RP/0/0/CPU0:XRV-8#traceroute mpls ipv4 2.2.2.2/32 source 8.8.8.8
Tracing MPLS Label Switched Path to 2.2.2.2/32, timeout is 2 seconds
  0 68.1.1.8 MRU 1500 [Labels: implicit-null/600004 Exp: 0/0]
L 1 68.1.1.6 MRU 1500 [Labels: implicit-null/400001 Exp: 0/0] 10 ms
L 2 46.1.1.4 MRU 1500 [Labels: implicit-null Exp: 0] 10 ms
! 3 24.1.1.2 10 ms
```

可以看到，节点 8 从 IBGP 邻居收到的外部路由标签信息的下一跳变更为

6.6.6.6，而 LDP 已经建立了关于 6.6.6.6 的有效标签转发信息，通过迭代就可以使得外部路由 MPLS 转发表有效。通过 MPLS traceroute 测试可以验证节点 8 与节点 2 之间已经建立了正确有效的 MPLS 转发路径。

6. BGP-LU 和 LDP 标签分配的一致性

BGP-LU 和 LDP 两种协议都会为前缀分配标签。对于不同的前缀可以分配不同的标签。但是，当两种协议为相同的前缀分配标签时，如果分配标签不一致就会引发数据面的转发问题。究竟 MPLS 数据面使用 BGP-LU 分配的标签还是 LDP 分配的标签呢。IOS 的解决方法非常简单，那就是让两个协议为相同的前缀分配同样的标签。

查看节点 8 的 LDP 和 BGP 为前缀 6.6.6.6/32 分配的标签情况：

```
RP/0/0/CPU0:XRV-8#show mpls ldp bindings 6.6.6.6/32
6.6.6.6/32, rev 13
          Local binding: label: 800000
          Remote bindings: (2 peers)
              Peer                    Label
              -----------------       -------
              6.6.6.6:0               ImpNull
              7.7.7.7:0               700003
-----------------------------------------------------------------
RP/0/0/CPU0:XRV-8#show bgp 6.6.6.6/32
Sun Oct 30 01:36:37.904 UTC
BGP routing table entry for 6.6.6.6/32
Versions:
  Process                 bRIB/RIB  SendTblVer
  Speaker                    23         23
    Local Label: 800000
Last Modified: Oct 30 00:50:07.739 for 00:46:30
Paths: (1 available, best #1)
  Not advertised to any peer
  Path #1: Received by speaker 0
  Not advertised to any peer
  Local
```

```
    6.6.6.6 (metric 2) from 6.6.6.6 (6.6.6.6)
      Received Label 3
      Origin IGP, metric 0, localpref 100, valid, internal,
best, group-best, labeled-unicast
      Received Path ID 0, Local Path ID 0, version 23
------------------------------------------------------------
RP/0/0/CPU0:XRV-8#show mpls forwarding
Sun Oct 30 01:39:18.793 UTC
Local    Outgoing   Prefix       Outgoing      Next Hop    Bytes
Label    Label      or ID        Interface                 Switched
------   --------   ----------   -----------   ---------   --------
800000   Pop        6.6.6.6/32   Gi0/0/0/6     68.1.1.6    4234
800001   Pop        7.7.7.7/32   Gi0/0/0/7     78.1.1.7    4272
800002   600004     2.2.2.2/32                 6.6.6.6     0
800003   600005     4.4.4.4/32                 6.6.6.6     0
800004   600006     5.5.5.5/32                 6.6.6.6     0
800005   Pop        67.1.1.0/24  Gi0/0/0/6     68.1.1.6    0
         Pop        67.1.1.0/24  Gi0/0/0/7     78.1.1.7    0
```

可以看到，BGP 和 LDP 进程均为相同的前缀分配了相同的标签信息，进而确保本地关于该前缀标签在 MPLS 转发表项中的一致性。

7.2.4 数据面验证

1．单径转发验证

前文已经验证了节点 8 和节点 2 之间 MPLS 转发路径，它们之间采用单层标签封装。节点 2 去往节点 8 的路径追踪结果如下：

```
RP/0/0/CPU0:XRV-2#traceroute mpls ipv4 8.8.8.8/32 source 2.2.2.2
  0 24.1.1.2 MRU 1500 [Labels: implicit-null/400008 Exp: 0/0]
L 1 24.1.1.4 MRU 1500 [Labels: implicit-null/600000 Exp: 0/0] 10 ms
L 2 46.1.1.6 MRU 1500 [Labels: implicit-null Exp: 0] 10 ms
! 3 68.1.1.8 10 ms
```

修改节点 2 和节点 4 之间的链路代价，将其由默认代价修改为 100，这使得节点 2 去往节点 4 的下一跳由 24.1.1.4 改变为 23.1.1.3，进一步影响了节点 2 去往节点 8 的路径。

```
RP/0/0/CPU0:XRV-2(config)#router ospf 1 area 0 interface Gi0/0/0/4
cost 100
```

```
RP/0/0/CPU0:XRV-2#show mpls forwarding
Sat Oct 29 16:05:46.762 UTC
Local    Outgoing    Prefix        Outgoing       Next Hop    Bytes
Label    Label       or ID         Interface                  Switched
------   --------    ----------    -----------    --------    --------
200000   200001      4.4.4.4/32    Gi0/0/0/3      23.1.1.3    88
200001   200003      45.1.1.0/24   Gi0/0/0/3      23.1.1.3    0
200002   Pop         3.3.3.3/32    Gi0/0/0/3      23.1.1.3    57544
200003   Pop         35.1.1.0/24   Gi0/0/0/3      23.1.1.3    0
200004   300000      5.5.5.5/32    Gi0/0/0/3      23.1.1.3    57205
200005   500009      6.6.6.6/32                   5.5.5.5     0
200006   500010      7.7.7.7/32                   5.5.5.5     0
200007   500011      8.8.8.8/32                   5.5.5.5     0
```

```
RP/0/0/CPU0:XRV-2#traceroute mpls ipv4 8.8.8.8/32 source 2.2.2.2
  0 23.1.1.2 MRU 1500 [Labels: 300000/500011 Exp: 0/0]
L 1 23.1.1.3 MRU 1500 [Labels: implicit-null/500011 Exp: 0/0] 10 ms
L 2 35.1.1.5 MRU 1500 [Labels: implicit-null/600000 Exp: 0/0] 0 ms
L 3 56.1.1.6 MRU 1500 [Labels: implicit-null Exp: 0] 0 ms
! 4 68.1.1.8 10 ms
```

可以看到，在修改了 OSPF 链路代价后，节点 2 去往 AS20 的路由由原来的节点 4 改变为节点 5。此时，节点 2 发向目的 IP 地址 8.8.8.8 的报文将采用路由迭代方式，先命中 8.8.8.8/32MPLS 转发表项，压入标签 500011，然后下一跳查表命中 5.5.5.5/32 的 MPLS 转发表项，压入标签 300000，于是节点 2 就发出了两层标签封装的 MPLS 报文，外层标签 300000 在节点 3 被剥离后交给节点 5，节点 5 将标签 500011 交换为 600000 后发给节点 6，节点 6 弹出标签还原为 IP 分组后送给目的节点 8。

2. 多径转发验证

IGP-SR 实验验证了 SR-MPLS 在域内的多径转发能力,通过启动 BGP 的多下一跳功能,也可以在域间实现多径数据转发。BGP 默认采用最优单径转发模式,可以通过下面指令改变 BGP 的默认路径行为:

```
router bgp <asn> address-family ipv4 unicast maximum-paths
[ibgp|ebgp] <2-64>
```

在节点 5 上启动 EBGP 多一跳功能,查看节点 5 的路由、MPLS 转发表项如下:

```
RP/0/0/CPU0:XRV-5#router bgp 10 address-family ipv4 unicast
maximum-paths ebgp 4
---------------------------------------------------------------------
Routing entry for 8.8.8.8/32
  Known via "bgp 10", distance 20, metric 0, [ei]-bgp, labeled unic
ast (3107)
  Tag 20, type external
  Routing Descriptor Blocks
    56.1.1.6, from 56.1.1.6, BGP external, BGP multi path
      Route metric is 0
    57.1.1.7, from 57.1.1.7, BGP external, BGP multi path
      Route metric is 0
    57.57.57.7, from 57.57.57.7, BGP external, BGP multi path
      Route metric is 0
  No advertising protos.
---------------------------------------------------------------------
RP/0/0/CPU0:XRV-5#show mpls forwarding prefix 8.8.8.8/32
Local    Outgoing    Prefix        Outgoing      Next Hop       Bytes
Label    Label       or ID         Interface                    Switched
------   ---------   -----------   -----------   -----------    --------
500011   600000      8.8.8.8/32    Gi0/0/0/6     56.1.1.6       808
         700010      8.8.8.8/32    Gi0/0/0/7     57.1.1.7       536
         700010      8.8.8.8/32    Gi0/0/0/0     57.57.57.7     536
```

可以看到,节点 5 已经围绕 3 个 EBGP 下一跳建立了等代价路由。

第 7 章 SR BGP 原理与实战

目前仅在节点 5 启动了多径，为了验证多径 MPLS 转发，保持节点 2 与节点 4 的 OSPF 链路代价为 100，确保节点 2 去往节点 8 的路径经过节点 5。节点 2 发送 MPLS Multipath 查询，结果如下：

```
RP/0/0/CPU0:XRV-2#traceroute mpls multipath ipv4 8.8.8.8/32
source 2.2.2.2 verbose
LLL!
Path 0 found,
 output interface GigabitEthernet0/0/0/3 nexthop 23.1.1.3
 source 2.2.2.2 destination 127.0.0.0
  0 2.2.2.2 23.1.1.3 MRU 1500 [Labels: 300006/500011 Exp: 0/0]
multipaths 0
L 1 23.1.1.3 35.1.1.5 MRU 1500 [Labels: implicit-null/500011
Exp: 0/0] ret code 8 multipaths 1
L 2 35.1.1.5 56.1.1.6 MRU 1500 [Labels: implicit-null/600000
Exp: 0/0] ret code 8 multipaths 3
L 3 56.1.1.6 68.1.1.8 MRU 1500 [Labels: implicit-null Exp: 0]
ret code 8 multipaths 1
! 4 68.1.1.8, ret code 3 multipaths 0
L!
Path 1 found,
 output interface GigabitEthernet0/0/0/3 nexthop 23.1.1.3
 source 2.2.2.2 destination 127.0.0.8
  0 2.2.2.2 23.1.1.3 MRU 1500 [Labels: 300006/500011 Exp: 0/0] mult
ipaths 0
L 1 23.1.1.3 35.1.1.5 MRU 1500 [Labels: implicit-null/500011
Exp: 0/0] ret code 8 multipaths 1
L 2 35.1.1.5 57.1.1.7 MRU 1500 [Labels: implicit-null/700010
Exp: 0/0] ret code 8 multipaths 3
L 3 57.1.1.7 78.1.1.8 MRU 1500 [Labels: implicit-null Exp: 0]
ret code 8 multipaths 1
! 4 78.1.1.8, ret code 3 multipaths 0
L!
Path 2 found,
```

```
output interface GigabitEthernet0/0/0/3 nexthop 23.1.1.3
source 2.2.2.2 destination 127.0.0.3
  0 2.2.2.2 23.1.1.3 MRU 1500 [Labels: 300006/500014 Exp: 0/0] mult
ipaths 0
L 1 23.1.1.3 35.1.1.5 MRU 1500 [Labels: implicit-null/500014
Exp: 0/0] ret code 8 multipaths 1
L 2 35.1.1.5 57.57.57.7 MRU 1500 [Labels: implicit-null/700010 Exp:
 0/0] ret code 8 multipaths 3
L 3 57.57.57.7 78.1.1.8 MRU 1500 [Labels: implicit-null Exp: 0]
ret code 8 multipaths 1
! 4 78.1.1.8, ret code 3 multipaths 0

Paths (found/broken/unexplored) (3/0/0)
Echo Request (sent/fail) (8/0)
Echo Reply (received/timeout) (8/0)
Total Time Elapsed 89 ms
```

探测发现，节点 2 到节点 8 存在 3 条路径，分别为节点 2→3→5→6→8、节点 2→3→5→7（链路 57.1.1.5-57.1.1.7）→8、节点 2→3→5→7（链路 57.57.57.5- 57.57.57.7）→8。

7.3 基于 BGP-LU 的 PrefixSID 分发

BGP 通过新定义的 BGP-PrefixSID 属性支持 BGP-SR 功能，通过实验掌握 BGP PrefixSID 的使能方式、PrefixSID 属性的构造、与 IGP-SR 的配合方式以及与 BGP-LU 分发标签的差异。

7.3.1 实验环境

采用图 7.5 拓扑进行实验，在第 7.2 节实验基础上，关闭 LDP 进程，并启动 OSPF-SR，让每个 OSPF 节点为本地 Loopback 接口地址分配 PrefixSID，将节点 2

和节点 4 链路代价恢复为 1，所有 BGP 节点配置不变。

7.3.2 BGP-LU 标签与 IGP-SR 标签的不一致性

按照实验环境设置要求，删除了所有节点的 LDP 配置，并在运行 OSPF 协议的节点上为各自 Loopback 接口手动分配 PrefixSID。节点 4 的基本配置如下：

```
route-policy SID($SID)
  set label-index $SID
end-policy
!
route-policy bgp_in
  pass
end-policy
!
route-policy bgp_out
  pass
end-policy
!
router static
 address-family ipv4 unicast
  46.1.1.6/32 GigabitEthernet0/0/0/6
 !
!
router ospf 1
 segment-routing mpls
 area 0
  interface Loopback0
   PrefixSID index 4
  !
  interface GigabitEthernet0/0/0/2
  !
  interface GigabitEthernet0/0/0/5
  !
```

```
!
!
router bgp 10
 address-family ipv4 unicast
  network 4.4.4.4/32
  allocate-label all
 !
 neighbor 2.2.2.2
  remote-as 10
  update-source Loopback0
  address-family ipv4 labeled-unicast
   route-policy bgp_in in
   route-policy bgp_out out
   next-hop-self
  !
 !
 neighbor 5.5.5.5
  remote-as 10
  update-source Loopback0
  address-family ipv4 labeled-unicast
   route-policy bgp_in in
   route-policy bgp_out out
   next-hop-self
  !
 !
 neighbor 46.1.1.6
  remote-as 20
  address-family ipv4 labeled-unicast
   route-policy bgp_in in
   route-policy bgp_out out
  !
 !
!
mpls oam
```

```
!
mpls label range table 0 400000 499999
```

首先，分析使用 OSPF-SR 代替 LDP 后，能否与 BGP-LU 配合实现 MPLS 数据转发。

保持 BGP-LU 配置不变，按照节点序号从小到大启动 OSPF 进程，发现在 OSPF 协议分发标签过程中会报告错误信息。例如，在节点 6 上启动 OSPF-SR 后，错误信息如下：

```
RP/0/0/CPU0:XRV-8#RP/0/0/CPU0: ipv4_rib[1148]: %ROUTING-RIB-3-
LABEL_ERR_ADD : Add local-label 16008 (2) for table 0xe0000000,
prefix 8.8.8.8/32, by proto ospf client 19 ospf node0_0_CPU0 -
existing label 600000 added by proto-id 6 client 21
--------------------------------------------------------------
RP/0/0/CPU0:XRV-8#show bgp 8.8.8.8/32
BGP routing table entry for 8.8.8.8/32
Versions:
  Process           bRIB/RIB     SendTblVer
  Speaker              22            22
    Local Label: 600000
Last Modified: Oct 30 00:50:07.687 for 01:33:44
Paths: (1 available, best #1)
  Advertised to update-groups (with more than one peer):
    0.1
  Path #1: Received by speaker 0
  Advertised to update-groups (with more than one peer):
    0.1
  Local
    8.8.8.8 (metric 2) from 8.8.8.8 (8.8.8.8)
      Received Label 3
      Origin IGP, metric 0, localpref 100, valid, internal,
best, group-best, labeled-unicast
      Received Path ID 0, Local Path ID 0, version 22
--------------------------------------------------------------
RP/0/0/CPU0:XRV-6#show mpls forwarding prefix 8.8.8.8/32
```

Local Label	Outgoing Label	Prefix or ID	Outgoing Interface	Next Hop	Bytes Switched
16008	Pop	SR Pfx (idx 8)	Gi0/0/0/8	68.1.1.8	59

报告错误的原因是本地 BGP 已经为 8.8.8.8/32 分配了标签 600000，而 OSPF 为该路由分配了另一个标签 16008。于是针对相同的前缀出现了标签不一致情况。回看第 7.2 节，BGP 和 LDP 会为相同的前缀自动分配相同标签，确保标签的一致性。而 OSPF 分配的标签是管理员从便于管理的角度出发手动设置的，这就与 BGP 自动分配的标签值产生了不一致的矛盾。但从最后 MPLS 转发表的成效情况来看，IOS 优选 OSPF 分配的标签信息生成 MPLS 转发表。

标签不一致的后果就是影响了 MPLS 数据转发，导致端到端通信路径的失效。在节点 2 上追踪去往节点 8 的 MPLS 路径，如下：

```
RP/0/0/CPU0:XRV-2#traceroute mpls ipv4 8.8.8.8/32 source 2.2.2.2

Tracing MPLS Label Switched Path to 8.8.8.8/32, timeout is 2 seconds

Codes: '!' - success, 'Q' - request not sent, '.' - timeout,
  'L' - labeled output interface, 'B' - unlabeled output interface,
  'D' - DS Map mismatch, 'F' - no FEC mapping, 'f' - FEC mismatch,
  'M' - malformed request, 'm' - unsupported tlvs, 'N' - no rx label,
  'P' - no rx intf label prot, 'p' - premature termination of LSP,
  'R' - transit router, 'I' - unknown upstream index,
  'X' - unknown return code, 'x' - return code 0

Type escape sequence to abort.

  0 24.1.1.2 MRU 1500 [Labels: implicit-null/400008 Exp: 0/0]
L 1 24.1.1.4 MRU 1500 [Labels: implicit-null/600000 Exp: 0/0] 0 ms
N 2 46.1.1.6 MRU 0 [No Label] 10 ms
```

可以看到，携带 400008 标签的 MPLS 分组在节点 4 被交换为 600000 后转发给节点 6，这是节点 6 的 BGP-LU 为 8.8.8.8/32 分配的 MPLS 标签，同时，节点 6 的 OSPF 为 8.8.8.8/32 分配了 16008 标签。对于节点 6 而言，MPLS 转发面认为 OSPF 分配的 16008 标签有效，而 BGP 分配的 600000 标签无效。于是，携带 600000 的 MPLS 到达节点 6 后会因为没有 MPLS 转发表项而被丢弃，故节点 2 发出路径探测消息显示了 "'N' - no rx label" 标识，即节点 6 没有对应的入标签转发表项。

通过上述实验验证了，BGP-LU 和 OSPF-SR 分配的标签不一致，会导致 MPLS 转发路径失效。为了解决这个问题，必须要确保 BGP 和 OSPF 分配的标签一致。

7.3.3 BGP–SR 的标签分配

为了保证 BGP 标签和 OSPF 标签的一致性，使用手动指定标签的方法。BGP 手动设置标签的方法通过 route-policy 命令实现，route-policy 设置标签索引的配置如下：

```
address-family ipv4 unicast
network <ip-address>/<mask-length> route-policy SID(<index >)
```

route-policy 可通过编程方式设置：

```
route-policy SID ($SID)
  set label-index $SID
end-policy
```

通过 route-policy 和 network 命令配合使用，就可人工为特定前缀指定 SID 索引号，即分配 BGP PrefixSID。在节点 8 上配置了 PrefixSID 后，在节点 4 上捕获到邻居节点 6 通告的 BGP-LU 更新报文如图 7.7 所示，可以看到在设置了 SID index 后，BGP-LU 通过 BGP PrefixSID 携带 SID index 信息，同时仍然通过 MP_REACH_NLRI 携带 LU 标签信息。从报文中携带的两个标签可以看出，BGP 节点会为同一个 BGP 前缀分配两个标签，一个为 BGP-LU 标签 60000，由节点 4 自己随机产生；另一个为 PrefixSID 8，对应标签值为 16008，由节点 8 生成、节点 6 中转的。这个 Update 报文被通告给节点 4。

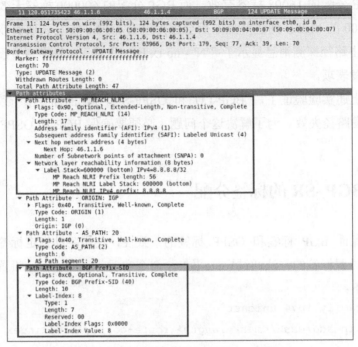

图 7.7 邻居节点 6 通告的 BGP-LU 更新报文

在节点 4 上查看其 MPLS 转发表项：

```
RP/0/0/CPU0:XRV-4#show mpls forwarding
Local    Outgoing    Prefix            Outgoing      Next Hop     Bytes
Label    Label       or ID             Interface                  Switched
------   --------    ---------------   -----------   ----------   --------
16002    Pop         SR Pfx (idx 2)    Gi0/0/0/2     24.1.1.2     69052
16003    16003       SR Pfx (idx 3)    Gi0/0/0/2     24.1.1.2     0
         16003       SR Pfx (idx 3)    Gi0/0/0/5     45.1.1.5     0
16005    Pop         SR Pfx (idx 5)    Gi0/0/0/5     45.1.1.5     2826747
400000   Pop         46.1.1.6/32       Gi0/0/0/6     46.1.1.6     122668
400006   Pop         6.6.6.6/32        Gi0/0/0/6     46.1.1.6     0
400007   600001      7.7.7.7/32        Gi0/0/0/6     46.1.1.6     0
400008   600000      8.8.8.8/32        Gi0/0/0/6     46.1.1.6     0
400009   Pop         SR Adj (idx 0)    Gi0/0/0/2     24.1.1.2     0
400010   Pop         SR Adj (idx 0)    Gi0/0/0/5     45.1.1.5     0
```

可以看出，节点 2 虽然收到邻居通告的关于 8.8.8.8/32 的两个标签，但是它选用了 BGP-LU 标签生成转发表，而不使用 BGP PrefixSID。也就是说即使手动配置 BGP PrefixSID，保证了与 OSPF 分配标签的一致性，但是，由于 BGP 在通告 BGP PrefixSID 时并行通告 BGP-LU 标签，而节点又优选了 BGP-LU 标签，最终导致转发面上 BGP 标签和 OSPF 标签不一致问题没有解决。

7.3.4 BGP-SR 标签与 BGP-LU 标签的一致性

需要改变 BGP 节点优选 PrefixSID 标签以解决标签不一致性问题。解决方法是禁止 BGP-LU 通告标签，或者禁止 BGP-LU 通告随机生成的标签并选用 BGP-SR 生成的标签，从而保证两个标签的一致性。从增量开发的角度来看，后者更具兼容性。

使能 BGP-SR 和 BGP-LU 分配相同标签的方法很简单，就是在节点上全局启动 Segment Routing 并且配置 SRGB。需要注意的是，全局启动 Segment Routing 功能要先于 BGP 进程启动，并且必须明确设定 SRGB 范围，否则配置就不会生效。在所有 BGP 节点上全局使能 SRGB 后，在节点 4 上捕获其发送的关于 8.8.8.8/32 的 BGP-LU 更新报文，如图 7.8 所示，可以看到此时 BGP-LU 标签和 BGP PrefixSID 保持一致。

图 7.8 携带一致标签信息的 BGP-LU 更新报文

在节点 4 上查看 BGP 信息和 MPLS 转发表项：

```
RP/0/0/CPU0:XRV-4#show bgp 8.8.8.8/32
BGP routing table entry for 8.8.8.8/32
Versions:
  Process           bRIB/RIB    SendTblVer
  Speaker                 8              8
    Local Label: 16008
Paths: (2 available, best #2)
  Advertised to update-groups (with more than one peer):
    0.3
  Path #1: Received by speaker 0
  Not advertised to any peer
  20
    5.5.5.5 (metric 2) from 5.5.5.5 (5.5.5.5)
      Received Label 16008
      Origin IGP, localpref 100, valid, internal, labeled-unicast
      Received Path ID 0, Local Path ID 0, version 0
      Prefix SID Attribute Size: 10
      Label Index: 8
  Path #2: Received by speaker 0
  Advertised to update-groups (with more than one peer):
    0.3
  20
    46.1.1.6 from 46.1.1.6 (6.6.6.6)
      Received Label 16008
      Origin IGP, localpref 100, valid, external, best, group-best, labeled-unicast
      Received Path ID 0, Local Path ID 0, version 8
      Origin-AS validity: not-found
      Prefix SID Attribute Size: 10
      Label Index: 8
-------------------------------------------------------------------
RP/0/0/CPU0:XRV-4#show mpls forwarding
Local   Outgoing    Prefix              Outgoing      Next Hop       Bytes
Label   Label       or ID               Interface                    Switched
```

```
-------  ---------  ------------------  ----------  ----------  --------
16002    Pop        SR Pfx (idx 2)      Gi0/0/0/2   24.1.1.2    6297
16003    16003      SR Pfx (idx 3)      Gi0/0/0/2   24.1.1.2    0
         16003      SR Pfx (idx 3)      Gi0/0/0/5   45.1.1.5    0
16005    Pop        SR Pfx (idx 5)      Gi0/0/0/5   45.1.1.5    2448
16006    Pop        SR Pfx (idx 6)      Gi0/0/0/6   46.1.1.6    0
16007    16007      SR Pfx (idx 7)      Gi0/0/0/6   46.1.1.6    0
16008    16008      SR Pfx (idx 8)      Gi0/0/0/6   46.1.1.6    0
400000   Pop        46.1.1.6/32         Gi0/0/0/6   46.1.1.6    3478
400009   Pop        SR Adj (idx 0)      Gi0/0/0/2   24.1.1.2    0
400010   Pop        SR Adj (idx 0)      Gi0/0/0/5   45.1.1.5    0
```

可以看到，节点 4 从节点 5 和节点 6 收到的 MPLS 标签均为 16008，而且本地也分配和使用 16008，从而实现了 BGP-LU 标签、BGP-SR 标签以及 OSPF-SR 标签三者的统一。

7.3.5 数据转发面验证

1. 单径转发验证

节点 2 去往节点 8 的路径追踪结果如下：

```
RP/0/0/CPU0:XRV-2#traceroute mpls ipv4 8.8.8.8/32 source 2.2.2.2
  0 24.1.1.2 MRU 1500 [Labels: implicit-null/16008 Exp: 0/0]
L 1 24.1.1.4 MRU 1500 [Labels: implicit-null/16008 Exp: 0/0] 10 ms
L 2 46.1.1.6 MRU 1500 [Labels: implicit-null Exp: 0] 10 ms
! 3 68.1.1.8 10 ms
```

可以修改节点 2 和节点 4 之间的链路代价，将其由默认值修改为 100，这样使得节点 2 去往节点 4 的下一跳由 24.1.1.4 改变为 23.1.1.3，这进一步影响了节点 2 去往节点 8 的路径。

```
RP/0/0/CPU0:XRV-2(config)#router ospf 1 area 0 interface Gi0/0/0/4
cost 100
```

```
RP/0/0/CPU0:XRV-2#show mpls forwarding
Local      Outgoing      Prefix              Outgoing    Next Hop    Bytes
```

```
Label      Label              or ID              Interface            Switched
------     ---------          ----------------   ----------           --------
16003      Pop                SR Pfx (idx 3)     Gi0/0/0/3 23.1.1.3   0
16004      16004              SR Pfx (idx 4)     Gi0/0/0/3 23.1.1.3   0
16005      16005              SR Pfx (idx 5)     Gi0/0/0/3 23.1.1.3   3661
16006      16006              SR Pfx (idx 6)               5.5.5.5    0
16007      16007              SR Pfx (idx 7)               5.5.5.5    0
16008      16008              SR Pfx (idx 8)               5.5.5.5    0
200008     Pop                SR Adj (idx 0)     Gi0/0/0/3 23.1.1.3   0
200009     Pop                SR Adj (idx 0)     Gi0/0/0/4 24.1.1.4   0
```

```
RP/0/0/CPU0:XRV-2#traceroute mpls ipv4 8.8.8.8/32 source 2.2.2.2
  0 23.1.1.2 MRU 1500 [Labels: 16005/16008 Exp: 0/0]
L 1 23.1.1.3 MRU 1500 [Labels: implicit-null/16008 Exp: 0/0] 10 ms
L 2 35.1.1.5 MRU 1500 [Labels: implicit-null/16008 Exp: 0/0] 10 ms
L 3 56.1.1.6 MRU 1500 [Labels: implicit-null Exp: 0] 10 ms
! 4 68.1.1.8 20 ms
```

可以看到，在修改了 OSPF 链路代价后，节点 2 去往 AS20 的路由由原来的节点 4 改变为节点 5。此时，节点 2 发向目的 IP 8.8.8.8 的报文将采用路由迭代方式，先命中 8.8.8.8/32 MPLS 转发表项，压入标签 16008，然后下一跳查表命中 5.5.5.5/32 的 MPLS 转发表项，压入标签 16005，于是节点 2 就发出了两层标签封装的 MPLS 报文，外层标签 16005 在节点 3 被剥离后交给节点 5，节点 5 将标签 16008 交换为 16008 后发给节点 6，节点 6 弹出标签并还原为 IP 分组后送给目的节点 8。

2. 多径转发验证

在节点 5 上启动 EBGP 多一跳功能，查看节点 5 的路由、MPLS 转发表项：

```
RP/0/0/CPU0:XRV-5#router bgp 10 address-family ipv4 unicast
maximum-paths ebgp 4
```

```
RP/0/0/CPU0:XRV-5#show ip route 8.8.8.8/32
Routing entry for 8.8.8.8/32
  Known via "bgp 10", distance 20, metric 0, [ei]-bgp, labeled unicast (3107), labeled SR
  Tag 20, type external
```

```
Installed Nov  3 03:00:35.140 for 01:46:07
Routing Descriptor Blocks
  56.1.1.6, from 56.1.1.6, BGP external, BGP multi path
    Route metric is 0
  57.1.1.7, from 57.1.1.7, BGP external, BGP multi path
    Route metric is 0
  57.57.57.7, from 57.57.57.7, BGP external, BGP multi path
    Route metric is 0
  No advertising protos.

RP/0/0/CPU0:XRV-5#show mpls forwarding prefix 8.8.8.8/32
Local   Outgoing    Prefix           Outgoing      Next Hop     Bytes
Label   Label       or ID            Interface                  Switched
------  ----------  ---------------  ------------  -----------  -----------
16008   16008       SR Pfx (idx 8)   Gi0/0/0/6     56.1.1.6     6112
        16008       SR Pfx (idx 8)   Gi0/0/0/7     57.1.1.7     3204
        16008       SR Pfx (idx 8)   Gi0/0/0/0     57.57.57.7   2948
```

可以看到节点 5 围绕 3 个 EBGP 下一跳建立了等代价路由。

目前仅在节点 5 启动了多径，为了验证多径 MPLS 转发，保持节点 2 与节点 4 的 OSPF 链路代价为 100，确保节点 2 去往节点 8 的路径经过节点 5。节点 2 发送 MPLS Multipath 查询，结果如下：

```
RP/0/0/CPU0:XRV-2#traceroute mpls multipath ipv4 8.8.8.8/32
source 2.2.2.2 verbose

LLL!
Path 0 found,
 output interface GigabitEthernet0/0/0/3 nexthop 23.1.1.3
 source 2.2.2.2 destination 127.0.0.0
  0 2.2.2.2 23.1.1.3 MRU 1500 [Labels: 16005/16008 Exp: 0/0]
multipaths 0
L 1 23.1.1.3 35.1.1.5 MRU 1500 [Labels: implicit-null/16008
Exp: 0/0] ret code 8 multipaths 1
```

```
L 2 35.1.1.5 56.1.1.6 MRU 1500 [Labels: implicit-null/16008
Exp: 0/0] ret code 8 multipaths 3
L 3 56.1.1.6 68.1.1.8 MRU 1500 [Labels: implicit-null Exp: 0]
ret code 8 multipaths 1
! 4 68.1.1.8, ret code 3 multipaths 0
L!
Path 1 found,
 output interface GigabitEthernet0/0/0/3 nexthop 23.1.1.3
 source 2.2.2.2 destination 127.0.0.8
   0 2.2.2.2 23.1.1.3 MRU 1500 [Labels: 16005/16008 Exp: 0/0] multipaths 0
L 1 23.1.1.3 35.1.1.5 MRU 1500 [Labels: implicit-null/16008
Exp: 0/0] ret code 8 multipaths 1
L 2 35.1.1.5 57.1.1.7 MRU 1500 [Labels: implicit-null/16008
Exp: 0/0] ret code 8 multipaths 3
L 3 57.1.1.7 78.1.1.8 MRU 1500 [Labels: implicit-null Exp: 0]
ret code 8 multipaths 1
! 4 78.1.1.8, ret code 3 multipaths 0
L!
Path 2 found,
 output interface GigabitEthernet0/0/0/3 nexthop 23.1.1.3
 source 2.2.2.2 destination 127.0.0.3
   0 2.2.2.2 23.1.1.3 MRU 1500 [Labels: 16005/16008 Exp: 0/0] multipaths 0
L 1 23.1.1.3 35.1.1.5 MRU 1500 [Labels: implicit-null/16008
Exp: 0/0] ret code 8 multipaths 1
L 2 35.1.1.5 57.57.57.7 MRU 1500 [Labels: implicit-null/16008
Exp: 0/0] ret code 8 multipaths 3
L 3 57.57.57.7 78.1.1.8 MRU 1500 [Labels: implicit-null Exp: 0]
ret code 8 multipaths 1
! 4 78.1.1.8, ret code 3 multipaths 0

 Paths (found/broken/unexplored) (3/0/0)
 Echo Request (sent/fail) (8/0)
 Echo Reply (received/timeout) (8/0)
 Total Time Elapsed 79 ms
```

第 7 章 SR BGP 原理与实战

探测发现，节点 2 到节点 8 存在 3 条路径，分别为节点 2→3→5→6→8、节点 2→3→5→7（链路 57.1.1.5-57.1.1.7）→8、节点 2→3→5→7（链路 57.57.57.5-57.57.57.7）→8。

7.4 BGP EPE

类似于 IGP 的 AdjSID，BGP 也可为 EBGP 邻居链路分配标签以实现精确的域间链路引流。BGP 定义了 Egress Peer Engineering，可以为邻居链路、邻居节点以及邻居集合分配标签。BGP 拓展了 BGP Link State 协议族实现 PeeringSID 信息的通告。

7.4.1 实验环境

拓展了图 7.5 实验拓扑，增加了节点 1，BGP EPE 实验拓扑如图 7.9 所示，各节点保持了第 7.3.1 节的配置。该拓扑中，节点 5 有 2 个 EBGP 对等节点，3 条邻居链路，节点 5 可以为节点 6 和节点 7 分配 PeerNodeSID、为节点 7 的 2 条物理链路分配 PeerAdjSID。新增节点 1 的目的是将其作为一个集中控制节点，从节点 5 收集 EPE 信息。

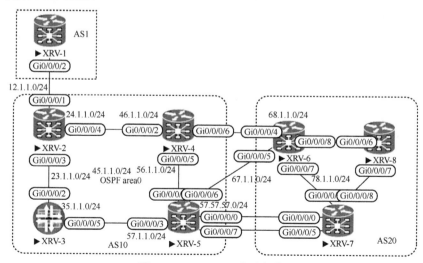

图 7.9 BGP EPE 实验拓扑

新增节点 1 的初始配置主要信息如下：

```
route-policy SID($SID)
  set label-index $SID
end-policy
!
route-policy bgp_in
  pass
end-policy
!
route-policy bgp_out
  pass
end-policy
!
router static
 address-family ipv4 unicast
  12.1.1.2/32 GigabitEthernet0/0/0/2
 !
!
router bgp 1
 address-family ipv4 unicast
  network 1.1.1.1/32 route-policy SID(1)
  allocate-label all
 !
 neighbor 12.1.1.2
  remote-as 10
  address-family ipv4 labeled-unicast
   route-policy bgp_in in
   route-policy bgp_out out
  !
 !
!
mpls oam
!
segment-routing
 global-block 16000 23999
```

7.4.2 EPE 配置与通告

1. PeeringSID 的分配机制

为邻居分配 PeeringSID 的指令为：

router bgp <asn> neighbor <ip-address> egress-engineering

为了实验验证 PeerNodeSID 和 PeerAdjSID 分配机制的差异性，先关闭节点 5 的 Gi0/0/0/6 接口，禁用与节点 6 的会话，仅保留与节点 7 的两条链路。为了描述方便，将节点 5 与 57.1.1.7 相连的链路称为 LowerLink，与 57.57.57.7 相连的链路称为 UpperLink。

在节点 5 针对邻居地址 57.1.1.7、57.57.57.7 两个 EBGP 会话配置"egress-engineering"参数，配置完成后查看本地分配的 EPE 标签信息：

```
RP/0/0/CPU0:XRV-5#show bgp egress-engineering

Egress Engineering Peer Set: 57.1.1.7/32 (12632f30)
    Nexthop: 57.1.1.7
    Version: 2084, rn_version: 2084
    Flags: 0x00000002
Local ASN: 10
Remote ASN: 20
Local RID: 5.5.5.5
Remote RID: 7.7.7.7
First Hop: 57.1.1.7
    NHID: 2
    IFH: 0x120
    Label: 500006, Refcount: 3
    rpc_set: 131ad9a4

Egress Engineering Peer Set: 57.57.57.7/32 (12632e8c)
    Nexthop: 57.57.57.7
    Version: 2029, rn_version: 2084
    Flags: 0x00000002
```

```
    Local ASN: 10
    Remote ASN: 20
    Local RID: 5.5.5.5
    Remote RID: 7.7.7.7
    First Hop: 57.57.57.7
        NHID: 3
        IFH: 0x40
        Label: 500007, Refcount: 3
    rpc_set: 131ac7ac

RP/0/0/CPU0:XRV-5#show mpls forwarding
Local    Outgoing    Prefix      Outgoing          Next Hop       Bytes
Label    Label       or ID       Interface                        Switched
------   --------    --------    -------------     ------------   --------
500006   Pop         No ID       Gi0/0/0/7         57.1.1.7       0
500007   Pop         No ID       Gi0/0/0/0         57.57.57.7     0
```

可以看出节点 5 为每个邻居链路分配标签信息并生成对应的 MPLS 转发表项信息。这两个标签类型均为 PeerNode 类型。当前这两个会话均是基于接口 IP 地址建立的 EBGP 会话，而不是基于 BGP RouterID。

下面在节点 5 和节点 7 之间基于 RouterID 建立 EBGP，进一步分析 PeerNode 和 PeerAdj 的生成机制。为此我们使用 5.5.5.5 和 7.7.7.7 地址建立 EBGP 会话，配置完成后查看本地分配 EPE 标签信息：

```
RP/0/0/CPU0:XRV-5#show bgp egress-engineering
Egress Engineering Peer Set: 7.7.7.7/32 (12632de8)
    Nexthop: 7.7.7.7
    Version: 2126, rn_version: 2127
    Flags: 0x00000006
    Local ASN: 10
    Remote ASN: 20
    Local RID: 5.5.5.5
    Remote RID: 7.7.7.7
    First Hop: 57.1.1.7
        NHID: 0
```

```
        IFH: 0x120
        Label: 500006, Refcount: 3
     rpc_set: 131ad9a4


  Egress Engineering Peer Set: 57.1.1.7/32 (12632f30)
     Nexthop: 57.1.1.7
     Version: 2127, rn_version: 2127
     Flags: 0x0000000a
  Local ASN: 10
  Remote ASN: 20
  Local RID: 5.5.5.5
  Remote RID: 7.7.7.7
  First Hop: 57.1.1.7
        NHID: 2
        IFH: 0x120
        Label: 500006, Refcount: 4
     rpc_set: 131ad9a4


  Egress Engineering Peer Set: 57.57.57.7/32 (12632e8c)
     Nexthop: 57.57.57.7
     Version: 2029, rn_version: 2127
     Flags: 0x00000002
  Local ASN: 10
  Remote ASN: 20
  Local RID: 5.5.5.5
  Remote RID: 7.7.7.7
  First Hop: 57.57.57.7
        NHID: 3
        IFH: 0x40
        Label: 500007, Refcount: 3
     rpc_set: 131ac7ac

------------------------------------------------------------------
RP/0/0/CPU0:XRV-5#show mpls forwarding
Local    Outgoing    Prefix         Outgoing      Next Hop      Bytes
```

```
Label      Label       or ID          Interface            Switched
------     --------    ---------      -----------          --------
500006     Pop         No ID          Gi0/0/0/7    57.1.1.7       0
500007     Pop         No ID          Gi0/0/0/0    57.57.57.7     0
```

可以看到 BGP 虽然为基于 RouterID 的会话分配了标签，但是这个标签并非新标签，而是复用了 LowerLink 的标签。从 MPLS 转发表项来看，其转发表项保持不变。究其原因是基于 RouterID 的会话的下一跳为 57.1.1.7，该会话所对应的链路实际上就是 LowerLink，因此没有必要为之分配新的标签。BGP 为基于 RouterID 的会话分配的标签类型仍然是 PeerNode。但这个操作对已有的 LowerLink 标签类型产生了影响，会促使 BGP 将 LowerLink 标签类型由 PeerNode 变更为 PeerAdj 类型。

当前情况下，虽然节点 5 和节点 7 之间存在两条链路，但是没有哪个标签能够将 UpperLink 和 LowerLink 同时使用起来，从而使得数据能够在两条链路上负载均衡。这与 BGP 的最优路径选择机制有关，默认情况下 BGP 仅使用一条路径进行数据转发。我们在节点 5 上开启多径路由能力，设置最大 EBGP 路径为 2，此时，本地分配 EPE 标签信息如下：

```
Egress Engineering Peer Set: 7.7.7.7/32 (12632de8)
    Nexthop: 7.7.7.7
    Version: 2698, rn_version: 2700
    Flags: 0x00000006
  Local ASN: 10
  Remote ASN: 20
  Local RID: 5.5.5.5
  Remote RID: 7.7.7.7
  First Hop: 57.1.1.7, 57.57.57.7
      NHID: 0, 0
      IFH: 0x120, 0x40
      Label: 500000, Refcount: 3
    rpc_set: 13563398

Egress Engineering Peer Set: 57.1.1.7/32 (12632f30)
    Nexthop: 57.1.1.7
```

```
   Version: 2700, rn_version: 2700
   Flags: 0x0000000a
  Local ASN: 10
  Remote ASN: 20
  Local RID: 5.5.5.5
  Remote RID: 7.7.7.7
  First Hop: 57.1.1.7
     NHID: 2
     IFH: 0x120
     Label: 500006, Refcount: 4
   rpc_set: 131ad9a4


 Egress Engineering Peer Set: 57.57.57.7/32 (12632e8c)
   Nexthop: 57.57.57.7
   Version: 2699, rn_version: 2700
   Flags: 0x0000000a
  Local ASN: 10
  Remote ASN: 20
  Local RID: 5.5.5.5
  Remote RID: 7.7.7.7
  First Hop: 57.57.57.7
     NHID: 3
     IFH: 0x40
     Label: 500007, Refcount: 4
   rpc_set: 131ac7ac

-----------------------------------------------------------------
RP/0/0/CPU0:XRV-5#show mpls forwarding
Local      Outgoing     Prefix       Outgoing        Next Hop      Bytes
Label      Label        or ID        Interface                     Switched
------     --------     ---------    ------------    -----------   --------
500000     Pop          No ID        Gi0/0/0/7       57.1.1.7      0
           Pop          No ID        Gi0/0/0/0       57.57.57.7    0
500006     Pop          No ID        Gi0/0/0/7       57.1.1.7      0
500007     Pop          No ID        Gi0/0/0/0       57.57.57.7    0
```

可以看到当邻居节点之间存在多条链路时，PeerNodeSID 与 PeerAdjSID 的区别就体现出来，前者可标识去往邻居节点的多条链路，后者唯一标识了与邻居节点的单条链路。使用 PeerNodeSID 引导流量时，在两个 BGP 节点之间的多条链路以负载均衡方式发送数据；使用 PeerAdjSID 则可以引导流量经过指定的链路。

2. PeeringSID 的通告机制

BGP 分配 PeeringSID 后需要通告出去，BGP 拓展了 MP_REACH_NLRI 属性的地址族来承载节点链路相关信息，并增加了 BGP-LS 属性承载相应的标签信息。

通常情况下，由跨域的集中式控制器收集域间链路的标签信息，以基于这些标签信息进行域间流量引导。在节点之间使能 BGP Link State 的配置方法如下。

首先，要启动 BGP Link State 地址族，指令如下：

```
router bgp <asn> address-family link-state link-state
```

然后，指定 BGP-LS 邻居，并设定 BGP 相关参数，主要指令包括：

```
neighbor a.b.c.d
  remote-as <asn>
  ebgp-multihop <1-255>
  update-source Loopback0
  address-family link-state link-state
   route-policy bgp_in in
   route-policy bgp_out out
```

BGP-LS 通常在 EBGP 邻居之间建立，如节点 1 和节点 5。两者既不属于同一个 AS，也不是直连 BGP 邻居。这与默认条件下的 EBGP 会话建立原则相违背。通常情况下，EBGP 对等体必须是直连的，它发出的 BGP 报文 IP TTL 设置为 1，确保 EBGP 报文不会跨越多个路由器传递。而在 BGP-LS 场景下，要跨越多个节点建立 EBGP 会话，此时需要通过"ebgp-multihop"指令进行修改。

在节点 1 和节点 5 完成 BGP-LS 配置后，查看节点 1 的 Link State 信息：

```
RP/0/0/CPU0:XRV-1#show bgp link-state link-state

Status codes: s suppressed, d damped, h history, * valid, > best
```

第 7 章 SR BGP 原理与实战

```
                      i - internal, r RIB-failure, S stale, N Nexthop-
discard
Origin codes: i - IGP, e - EGP, ? - incomplete
Prefix codes: E link, V node, T IP reacheable route, u/U unknown
              I Identifier, N local node, R remote node, L link, P
prefix
              L1/L2 ISIS level-1/level-2, O OSPF, D direct, S
static/peer-node
              a area-ID, l link-ID, t topology-ID, s ISO-ID,
              c confed-ID/ASN, b bgp-identifier, r router-ID,
              i if-address, n nbr-address, o OSPF Route-type, p
 IP-prefix
              d designated router address
   Network            Next Hop         Metric LocPrf Weight Path
*> [E][B][I0x0][N[c10][b0.0.0.0][q5.5.5.5]][R[c20][b0.0.0.0][q7.7.7.7]]
[L[i5.5.5.5][n7.7.7.7]]/664
                     5.5.5.5                              0 10 i
*> [E][B][I0x0][N[c10][b0.0.0.0][q5.5.5.5]][R[c20][b0.0.0.0][q7.7.7.7]]
[L[i57.1.1.5][n57.1.1.7]]/664
                     5.5.5.5                              0 10 i
*> [E][B][I0x0][N[c10][b0.0.0.0][q5.5.5.5]][R[c20][b0.0.0.0][q7.7.7.7]]
[L[i57.57.57.5][n57.57.57.7]]/664
                     5.5.5.5                              0 10 i
Processed 3 prefixes, 3 paths
```

通过 Link State 的邻居状态看到的信息比较混乱，还有一种更为直观地查看链路状态信息的方法，就是启动 PCE 进程，指令如下：

```
pce address (ipv4 | ipv6) <ip-address>
```

重新查看 EPE 信息，如下：

```
RP/0/0/CPU0:XRV-1#show pce ipv4 topology

PCE's topology database - detail:
----------------------------------
Node 1
```

```
BGP router ID: 5.5.5.5 ASN: 10

Link[0]: local address 5.5.5.5, remote address 7.7.7.7
  Local node:
    BGP router ID: 5.5.5.5 ASN: 10
  Remote node:
    BGP router ID: 7.7.7.7 ASN: 20
  Metric: IGP 0, TE 0
  Bandwidth: Total 0, Reservable 0
  Adj SID: 500000 (epe)

Link[1]: local address 57.1.1.5, remote address 57.1.1.7
  Local node:
    BGP router ID: 5.5.5.5 ASN: 10
  Remote node:
    BGP router ID: 7.7.7.7 ASN: 20
  Metric: IGP 0, TE 0
  Bandwidth: Total 0, Reservable 0
  Adj SID: 500006 (epe)

Link[2]: local address 57.57.57.5, remote address 57.57.57.7
  Local node:
    BGP router ID: 5.5.5.5 ASN: 10
  Remote node:
    BGP router ID: 7.7.7.7 ASN: 20
  Metric: IGP 0, TE 0
  Bandwidth: Total 0, Reservable 0
  Adj SID: 500007 (epe)

Node 2
  BGP router ID: 7.7.7.7 ASN: 20
```

7.5 混合多自治域环境下的混合 SID 传递实验

通过前序章节的阐述，已经讲解了 IGP、BGP 分发节点和链路的 SID 信息，并建立基于尽力服务的路径转发机制。进一步实验了通过路径编排方法，基于 SID 实现流量的引导。

实际运行场景比较复杂，往往部署域内协议和域间协议，通过协议发布和交互实现跨自制域之间的互联。在这种场景下，SR 技术提供了一种更为灵活的路径控制方式。试想如果存在一个集中路径控制节点，能够掌握由域内、域间的节点和链路组成的完整的拓扑信息，并在感知到所有节点和链路带宽、时延以及负载等参数情况下，就可以基于端到端数据转发需求，按照一定的算法为之规划一条满足 QoS 需求的路径。这条路径实际上就是一个 SID 列表。集中路径控制节点通过策略将这条 SID 列表下发给流量的入口节点，其将满足给定条件的流量压入对应的 SID 列表，进而牵引流量按照预先规划的 QoS 路径到达目的地。路径的编排计算不需要网络中路由节点的参与，网络节点也不需要维护 QoS 路径的信息，整个机制对 QoS 支撑都是无状态的，因而具有很好的扩展性。这种方法需要集中路径控制节点实时感知整个网络节点的负载情况，并依据这些参数的动态变化实时调整路径，这给路径的实时编排和动态更新带来了一定挑战。

7.5.1 实验环境

在图 7.9 拓扑上增加节点 9，其属于 AS9，混合 SID 传递实验拓扑如图 7.10 所示。节点 3 仅运行 OSPF-SR，节点 1、节点 9 仅运行 BGP-SR，其他节点运行 BGP-SR 和 OSPF-SR。指令采用前序章节阐述的缺省配置方式。

启动 BGP 和 IGP 的节点，IGP 和 BGP 都为自己 Loopback 接口地址分配 IP 地址，保证 BGP 和 IGP 分配的 PrefixSID 相同。

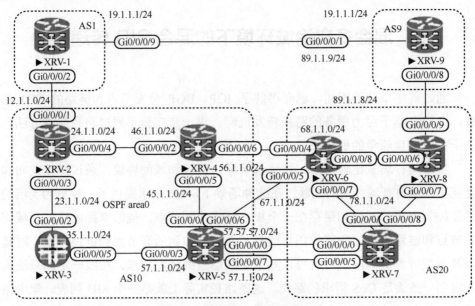

图 7.10 混合 SID 传递实验拓扑

下面从节点 1 的视角，以节点 9 的 Loopback 地址为目标，通过混合 SID 方式阐述 SR 控制数据转发路径的灵活性。

7.5.2 缺省策略条件下的路由与数据转发

在默认 BGP 策略下按照 AS-PATH 的长短，优选最小 AS 跳数的邻居作为下一跳，因此，节点 1 去往节点 9 的 Loopback 接口地址 9.9.9.9/32 的下一跳即节点 9 与节点 1 互联的接口 IP 地址 19.1.1.9。

7.5.3 基于 BGP PrefixSID 实现跨 AS 引流

如果期望 AS1～AS9 的数据转发路径经过 AS10 和 AS20，则可以使用这两个 AS 中某个 AS 通告 PrefixSID 作为索引来引导流量。我们可以通过 16006 先将流量引导至节点 6 来达到此目的，使得按照节点 1→2→4→6→8→9 路径进行数据转发，采用 Nil-FEC 方式进行路径追踪，结果如下：

```
RP/0/0/CPU0:XRV-1#traceroute mpls nil-fec labels 16006,16009
source 1.1.1.1 output interface g0/0/0/2 nexthop 12.1.1.2

  0 12.1.1.1 MRU 1500 [Labels: 16006/16009/explicit-null Exp: 0/0/0]
L 1 12.1.1.2 MRU 1500 [Labels: implicit-null/16006/16009/explicit-
null Exp: 0/0/0/0] 0 ms
L 2 24.1.1.4 MRU 1500 [Labels: implicit-null/implicit-null/16009/
explicit-null Exp: 0/0/0/0] 10 ms
L 3 46.1.1.6 MRU 1500 [Labels: implicit-null/16009/explicit-null
Exp: 0/0/0] 10 ms
L 4 68.1.1.8 MRU 1500 [Labels: implicit-null/implicit-null/explicit-
null Exp: 0/0/0] 10 ms
! 5 89.1.1.9 10 ms
```

或者通过 16004 将流量引导至节点 4 来达到目的,此时依次经过节点 1→2→4→5→6→8→9,路径追踪结果如下:

```
RP/0/0/CPU0:XRV-1#traceroute mpls nil-fec labels 16005,16009
source 1.1.1.1 output interface g0/0/0/2 nexthop 12.1.1.2

  0 12.1.1.1 MRU 1500 [Labels: 16005/16009/explicit-null Exp: 0/0/0]
L 1 12.1.1.2 MRU 1500 [Labels: 16005/16009/explicit-null Exp: 0/0/0]
0 ms
L 2 24.1.1.4 MRU 1500 [Labels: implicit-null/16009/explicit-null
Exp: 0/0/0] 10 ms
L 3 45.1.1.5 MRU 1500 [Labels: implicit-null/16009/explicit-null
Exp: 0/0/0] 10 ms
L 4 56.1.1.6 MRU 1500 [Labels: implicit-null/16009/explicit-null
Exp: 0/0/0] 20 ms
L 5 68.1.1.8 MRU 1500 [Labels: implicit-null/implicit-null/
explicit-null Exp: 0/0/0] 10 ms
! 6 89.1.1.9 20 ms
```

7.5.4 基于 BGP PeeringSID 实现跨域间链路引流

如果期望控制 AS10 和 AS20 之间的域间路径,仅仅靠 BGP PrefixSID 是无法完成的。这个时候就需要用到 BGP PeeringSID。节点 5 去往节点 9 的最优下一跳为节点 6,如果期望走节点 6 和节点 7 之间的 LowerLink(57.1.1.0/24),则需要将 EPE 标签 500006 编排进去:

```
RP/0/0/CPU0:XRV-1#traceroute mpls nil-fec labels 16005,500006,
16009 output interface Gi0/0/0/2 nexthop 12.1.1.2 source 1.1.1.1

Tracing MPLS Label Switched Path with Nil FEC with labels
[16005,24007,16009], timeout is 2 seconds

 0 12.1.1.1 MRU 1500 [Labels: 16005/500006/16009/explicit-null
Exp: 0/0/0/0]
L 1 12.1.1.2 MRU 1500 [Labels: 16005/500006/16009/explicit-null
Exp: 0/0/0/0] 0 m          s
L 2 24.1.1.4 MRU 1500 [Labels: implicit-null/500006/16009/explicit-
null Exp: 0/0/          0/0] 10 ms
L 3 45.1.1.5 MRU 1500 [Labels: implicit-null/16009/explicit-null
Exp: 0/0/0] 20          ms
L 4 57.1.1.7 MRU 1500 [Labels: implicit-null/16009/explicit-null
Exp: 0/0/0] 20          ms
L 5 78.1.1.8 MRU 1500 [Labels: implicit-null/implicit-null/explicit-
null Exp: 0/          0/0] 20 ms
! 6 89.1.1.9 10 ms
```

7.5.5 基于 IGP SID 实现跨域内引流

通过上述方法可以控制流量经过的域间链路和节点,如果期望能够进一步控制经过每一个域内的链路,就需要知道 IGP 分配的 PrefixSID 和 AdjSID 信息。例如,AS10 中的节点 3 并没有运行 BGP-SR,所以节点 3 的 SID 并不能被节点 1 学习到，而且 IGP 分配的 AdjSID 信息也不会传播出去，这样就无法明确引导流量经过节点 3。

此时 BGP-LS 的作用就体现出来。BGP-LS 除了承载 BGP EPE PeeringSID 信息，还可以承载 IGP 的 Link-State 信息。

通常每个 AS 会部署路由反射器（Route Reflector，RR）提升 BGP 之间的消息交互效率，那么实施 SR Policy 的节点通常跟每个 AS 的 RR 互联即可。在 RR 上将 IGP Link State 发布到 BGP 中，并启动 BGP-LS，这个 BGP-LS 定向连接到 SR Policy 集中控制节点，进而收集到所有域内、域间链路和 PrefixSID 信息。

本实验将节点 1 复用为集中式 SR Policy 控制节点，假定 AS1 节点 5 和 AS20 节点 8 为该域的 RR，分别与节点 1 建立 BGP-LS，实验过程如下。

首先，将 OSPF 收集的拓扑信息发布到 BGP-LS 中，指令为：

```
router ospf <pid> distribute link-state
```

然后，让节点 5、节点 8 与节点 1 建立 BGP-LS 邻居会话，用于交互 BGP-LS 信息。节点 5 的相关配置信息为：

```
router ospf 1
 distribute link-state
 segment-routing mpls
 area 0
  interface Loopback0
   PrefixSID index 5
  !
  interface GigabitEthernet0/0/0/3
  !
  interface GigabitEthernet0/0/0/4
  !
 !
!
router bgp 10
 address-family ipv4 unicast
  maximum-paths ebgp 2
  network 5.5.5.5/32 route-policy SID(5)
  allocate-label all
```

```
!
address-family link-state link-state
!
neighbor 1.1.1.1
 remote-as 1
 ebgp-multihop 32
 update-source Loopback0
 ignore-connected-check
 address-family link-state link-state
  route-policy bgp_in in
  route-policy bgp_out out
 !
!
neighbor 2.2.2.2
 remote-as 10
 update-source Loopback0
 address-family ipv4 labeled-unicast
  route-policy bgp_in in
  route-policy bgp_out out
  next-hop-self
 !
!
neighbor 4.4.4.4
 remote-as 10
 update-source Loopback0
 address-family ipv4 labeled-unicast
  route-policy bgp_in in
  route-policy bgp_out out
  next-hop-self
 !
!
neighbor 7.7.7.7
 remote-as 20
 ebgp-multihop 64
```

```
   egress-engineering
   update-source Loopback0
   address-family ipv4 unicast
   !
   address-family ipv4 labeled-unicast
    route-policy bgp_in in
    route-policy bgp_out out
   !
  !
  neighbor 56.1.1.6
   remote-as 20
   egress-engineering
   address-family ipv4 labeled-unicast
    route-policy bgp_in in
    route-policy bgp_out out
   !
  !
  neighbor 57.1.1.7
   remote-as 20
   egress-engineering
   address-family ipv4 labeled-unicast
    route-policy bgp_in in
    route-policy bgp_out out
   !
  !
  neighbor 57.57.57.7
   remote-as 20
   egress-engineering
   address-family ipv4 labeled-unicast
    route-policy bgp_in in
    route-policy bgp_out out
   !
  !
```

```
mpls oam
!
segment-routing
 global-block 16000 23999
!
end
```

采用类似指令在节点 1 和节点 8 之间建立 BGP-LS 会话，使得节点 1 能够获得 IGP Link State 和 BGP EPE Link State 信息，并启动 PCE 进程以查看拓扑信息。

节点 1 上可以通过 BGP-LS 查看每一个域的拓扑信息：

```
RP/0/0/CPU0:XRV-1#show pce ipv4 topology 5.5.5.5
PCE's topology database - detail:
---------------------------------
Node 4
  OSPF router ID: 5.5.5.5 area ID: 0 ASN: 10
  BGP router ID: 5.5.5.5 ASN: 10
  Prefix SID:
    Prefix 5.5.5.5, label 16005 (regular)
  SRGB INFO:
    OSPF router ID: 5.5.5.5 area ID: 0 SRGB Start: 16000 Size: 8000

  Link[0]: local address 35.1.1.5, remote address 35.1.1.3
    Local node:
      OSPF router ID: 5.5.5.5 area ID: 0 ASN: 10
    Remote node:
      OSPF router ID: 3.3.3.3 area ID: 0 ASN: 10
    Metric: IGP 1, TE 1
    Bandwidth: Total 0, Reservable 0
    Adj SID: 500001 (unprotected)

  Link[1]: local address 45.1.1.5, remote address 45.1.1.4
    Local node:
      OSPF router ID: 5.5.5.5 area ID: 0 ASN: 10
    Remote node:
```

```
        OSPF router ID: 4.4.4.4 area ID: 0 ASN: 10
      Metric: IGP 1, TE 1
      Bandwidth: Total 0, Reservable 0
      Adj SID: 500000 (unprotected)

  Link[2]: local address 56.1.1.5, remote address 56.1.1.6
    Local node:
        BGP router ID: 5.5.5.5 ASN: 10
    Remote node:
        BGP router ID: 6.6.6.6 ASN: 20
      Metric: IGP 0, TE 0
      Bandwidth: Total 0, Reservable 0
      Adj SID: 500005 (epe)

  Link[3]: local address 57.1.1.5, remote address 57.1.1.7
    Local node:
        BGP router ID: 5.5.5.5 ASN: 10
    Remote node:
        BGP router ID: 7.7.7.7 ASN: 20
      Metric: IGP 0, TE 0
      Bandwidth: Total 0, Reservable 0
      Adj SID: 500006 (epe)

  Link[4]: local address 57.57.57.5, remote address 57.57.57.7
    Local node:
        BGP router ID: 5.5.5.5 ASN: 10
    Remote node:
        BGP router ID: 7.7.7.7 ASN: 20
      Metric: IGP 0, TE 0
      Bandwidth: Total 0, Reservable 0
      Adj SID: 500007 (epe)
RP/0/0/CPU0:XRV-1#show pce ipv4 topology 8.8.8.8

PCE's topology database - detail:
```

```
----------------------------------
Node 7
  OSPF router ID: 8.8.8.8 area ID: 0 ASN: 20
  Prefix SID:
    Prefix 8.8.8.8, label 16008 (regular)
  SRGB INFO:
    OSPF router ID: 8.8.8.8 area ID: 0 SRGB Start: 16000 Size: 8000

  Link[0]: local address 68.1.1.8, remote address 68.1.1.6
    Local node:
      OSPF router ID: 8.8.8.8 area ID: 0 ASN: 20
    Remote node:
      OSPF router ID: 6.6.6.6 area ID: 0 ASN: 20
    Metric: IGP 1, TE 1
    Bandwidth: Total 0, Reservable 0
    Adj SID: 800002 (unprotected)

  Link[1]: local address 78.1.1.8, remote address 78.1.1.7
    Local node:
      OSPF router ID: 8.8.8.8 area ID: 0 ASN: 20
    Remote node:
      OSPF router ID: 7.7.7.7 area ID: 0 ASN: 20
    Metric: IGP 1, TE 1
    Bandwidth: Total 0, Reservable 0
    Adj SID: 800000 (unprotected)
```

为了简化篇幅，我们仅在节点1查看了节点5和节点8的链路状态信息，可以看到节点5和节点8的PrefixSID以及BGP、IGP链路SID信息均上报给了节点1。

为了让流量经过节点 2→3 链路，需要知道节点 2 与节点 3 互联链路的AdjSID信息，查看节点2的链路状态信息：

```
RP/0/0/CPU0:XRV-1#show pce ipv4 topology 2.2.2.2
PCE's topology database - detail:
----------------------------------
```

```
Node 3
  OSPF router ID: 2.2.2.2 area ID: 0 ASN: 10
  Prefix SID:
    Prefix 2.2.2.2, label 16002 (regular)
  SRGB INFO:
    OSPF router ID: 2.2.2.2 area ID: 0 SRGB Start: 16000 Size: 8000

  Link[0]: local address 23.1.1.2, remote address 23.1.1.3
    Local node:
      OSPF router ID: 2.2.2.2 area ID: 0 ASN: 10
    Remote node:
      OSPF router ID: 3.3.3.3 area ID: 0 ASN: 10
    Metric: IGP 1, TE 1
    Bandwidth: Total 0, Reservable 0
    Adj SID: 200001 (unprotected)

  Link[1]: local address 24.1.1.2, remote address 24.1.1.4
    Local node:
      OSPF router ID: 2.2.2.2 area ID: 0 ASN: 10
    Remote node:
      OSPF router ID: 4.4.4.4 area ID: 0 ASN: 10
    Metric: IGP 1, TE 1
    Bandwidth: Total 0, Reservable 0
    Adj SID: 200000 (unprotected)         Bandwidth: Total 0,
Reservable 0
```

可以看到节点 2 为节点 2→3 链路分配了 200001 标签。

综合上述标签信息，编辑 SID 列表为 200001,16005, 500006, 16006,16009 来引导去往节点 9 的流量，路径追踪结果如下：

```
RP/0/0/CPU0:XRV-1#traceroute mpls nil-fec labels 200001,16005,
500006,16006,16009 output interface Gi0/0/0/2 nexthop 12.1.1.2
source 1.1.1.1
```

```
Tracing MPLS Label Switched Path with Nil FEC with labels
[24000,16005,24003,16006,16009], timeout is 2 seconds
24000

  0 12.1.1.1 MRU 1500 [Labels: 200001/16005/500006/16006/16009/
explicit-null Exp: 0/0/0/0/0/0]
L 1 12.1.1.2 MRU 1500 [Labels: implicit-null/16005/500006/16006/
16009/explicit-null Exp: 0/0/0/0/0/0] 10 ms
. 2 *
L 3 35.1.1.5 MRU 1500 [Labels: implicit-null Exp: 0] 10 ms
L 4 57.1.1.7 MRU 1500 [Labels: implicit-null Exp: 0] 20 ms
L 5 67.1.1.6 MRU 1500 [Labels: implicit-null/16009 Exp: 0/0] 20 ms
L 6 68.1.1.8 MRU 1500 [Labels: implicit-null/implicit-null Exp:
0/0] 20 ms
! 7 89.1.1.9 10 ms
```

可以看到，节点 1 发出的 MPLS Echo 报文按照指定的路径依次经过节点 1→2→3→5→7→6→8→9。

需要特殊说明的是*号行：当节点 1 发出 TTL=2 的报文时，达到节点 3 时出现 TTL 超时，节点生成目的 IP 为 1.1.1.1 的 MPLS Echo Reply 报文，查看节点 3 的路由和 MPLS 转发表信息：

```
RP/0/0/CPU0:XRV-3#show ip route


O    2.2.2.2/32 [110/2] via 23.1.1.2, 1d00h, GigabitEthernet0/0/0/2
L    3.3.3.3/32 is directly connected, 1d03h, Loopback0
O    4.4.4.4/32 [110/3] via 23.1.1.2, 23:59:10, GigabitEthernet0/0/0/2
                [110/3] via 35.1.1.5, 23:59:10, GigabitEthernet0/0/0/5
O    5.5.5.5/32 [110/2] via 35.1.1.5, 23:59:10, GigabitEthernet0/0/0/5
C    23.1.1.0/24 is directly connected, 1d03h, GigabitEthernet0/0/0/2
L    23.1.1.3/32 is directly connected, 1d03h, GigabitEthernet0/0/0/2
O    24.1.1.0/24 [110/2] via 23.1.1.2, 1d00h, GigabitEthernet0/0/0/2
C    35.1.1.0/24 is directly connected, 1d03h, GigabitEthernet0/0/0/5
L    35.1.1.3/32 is directly connected, 1d03h, GigabitEthernet0/0/0/5
O    45.1.1.0/24 [110/2] via 35.1.1.5, 23:59:10, GigabitEthernet0/0/0/5
```

```
L    127.0.0.0/8 [0/0] via 0.0.0.0, 1d03h
--------------------------------------------------------------
RP/0/0/CPU0:XRV-3#show mpls forwarding
Local   Outgoing    Prefix            Outgoing       Next Hop      Bytes
Label   Label       or ID             Interface                    Switched
------  ----------  ----------------  -------------  ------------  --------
16002   Pop         SR Pfx (idx 2)    Gi0/0/0/2      23.1.1.2      416
16004   16004       SR Pfx (idx 4)    Gi0/0/0/2      23.1.1.2      0
        16004       SR Pfx (idx 4)    Gi0/0/0/5      35.1.1.5      0
16005   Pop         SR Pfx (idx 5)    Gi0/0/0/5      35.1.1.5      68483
300000  Pop         SR Adj (idx 0)    Gi0/0/0/5      35.1.1.5      0
300001  Pop         SR Adj (idx 0)    Gi0/0/0/2      23.1.1.2      0
```

由于节点 3 仅运行 OSPF 协议，只建立 IGP 路由信息，并没有 AS 外部路由信息，而该 Echo Reply 报文的目的 IP 为 1.1.1.1，既无法命中 MPLS 转发表项，也没有命中路由表，因此节点 3 将丢弃该 Echo Reply 分组，导致节点 1 发出的 TTL=2 Request 报文得不到响应。

第 8 章
基于 SR 的 FRR 技术原理与实战

8.1 FRR 技术原理

通过路由协议自身机制发现故障，扩散故障信息，重新计算收敛路径，整个过程所需操作耗时很长，会对丢包和时延敏感类业务产生较大影响。原因包括两个方面，一是故障发生时，数据到故障点后因为路由的失效，导致数据丢弃；二是协议收敛时，各个节点之间持有的拓扑信息不一致，路由计算快慢不同，路由生效时间也不一样，造成了协议收敛期间数据转发出现短暂的环路，即微环路。

快速重路由（Fast ReRoute，FRR）技术用于解决链路故障发生时路由黑洞和微环路期间的数据丢弃问题，是对丢包敏感类等 QoS 要求较高的业务的有效支撑手段。

快速重路由主要包含以下 3 个方面的内容。

1. 快速故障检测

双向转发检测（Bidirectional Forwarding Detection，BFD）技术专门用于检测邻居之间的互通性故障，通过双向或者单向快速发送探测报文，能够在毫秒级时间内检测出故障问题。

2. 保护路径机制

当故障点的上游节点检测到故障后，立刻调用预先计算的保护路径，保证数据的持续传送，并且确保这条保护路径不被接下来协议的收敛过程所影响。

保护路径示意图如图 8.1 所示，节点 S 针对节点 S→F 链路预先计算保护路径 T(S→Q)，在检测到链路故障后直接将去往目的节点 D 的流量导入保护路径，被保护的流量在节点 Q 点释放后，再沿着正常路径转发到目的地，确保数据的持续转发。其中保护路径 T(S→Q)通过预先计算，用以引导流量绕过故障点并最终送达目的地。故障点的上游节点是预先计算保护路径并实施重路由的节点，这个节点常被称为本地修复节点（Point of Local Repair，PLR）。

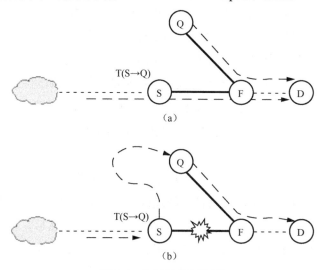

图 8.1　保护路径示意图

3．无环路收敛机制

网络节点发现拓扑变化后并不立刻采用新路由，而是持续使用当前路由一段时间；与此同时，计算一条不受拓扑变化影响的无环保护路径；并使用无环保护路径一段时间；在确保整个网络收敛后，从无环保护路径切换到收敛路径。

保护路径机制和无环路收敛机制主要用于解决网络拓扑变化时短暂的数据转发失效问题，但是两者侧重点各有不同，前者侧重于网络故障时的快速保护，即将流量从正常转发路径迁移到保护路径上，保障数据的持续转发；后者则着重解决协议收敛期间，在各个节点路由不一致的情况下，避免路由环路导致的数据转发环路问题，保障数据能够有效转发。从路径计算上来说，前者针对特定保护对象预先计算路径；后者则是在获知拓扑发生变化后，实时计算保护路径。保护路径机制用于解决故障发生时的问题，无环路收敛机制用于解决收敛过程中的问题，将上述两种机制融合是实现网络事件发生后、协议收敛过程中数据转发不中断的有效方案。

8.2 保护路径计算与使用

快速重路由技术是节点本地的一种保护路径计算技术，算法基于已有的路由协议和标签分发协议获得的信息，按照厂商独有的计算算法实现保护路径的计算，生成保护路径，支撑主用路径失效后的快速重路由。因此，快速重路由技术无须专用协议的支撑，它只是利用了已有的协议和数据转发机制实现针对特定前缀或者链路的保护，网络上的节点不会感知到其他节点是否启用了快速重路由技术。

FRR技术的路径计算核心就是PQ节点的计算方法，下面对这几个基本概念简要阐述。

P空间：一组节点的集合，在给定组件故障发生前，PLR节点S就存在到达该集合内任意节点的路径，且路径不经过故障组件。对于PLR节点S而言，其P空间的计算方法为：首先计算PLR节点S的生成树，然后，从生成树上移除所有经过故障组件的节点，剩下的节点集合就构成了PLR节点S的P空间。

Q空间：一组节点的集合，集合内的节点到达目的节点的路径与给定的故障组件无关。针对被保护组件的Q空间及计算方法为：首先计算以目的节点D为根的逆向生成树，记为rSPT(D)。然后，从rSPT(D)上剪枝掉那些经过被保护组件的节点，剩下的节点集合构成Q空间。注意rSPT(D)使用指向根节点D方向的链路代价进行最短路径计算。

PQ节点：P空间与Q空间两个节点集合的交集。

如果PLR的邻居节点就是PQ节点，则可以在网络故障发生后，将流量导入邻居节点实现快速重路由，这种方法被称为无环备选（Loop-Free Alternate，LFA）路径技术。

如果PLR的邻居节点都不是PQ节点，那么PLR就无法通过邻居节点实现快速重路由；如果此时网络中存在PQ节点，那么PLR可与PQ节点建立一条隧道，当网络故障发生后，通过隧道将流量引导至PQ节点，进而实现快速重路由，这种方法被称为远端无环备选（Remote Loop-Free Alternate，R-LFA）路径技术。

当网络因为拓扑原因导致不存在 PQ 节点时，LFA 路径和 R-LFA 路径无法工作。但只要当前节点与目的节点之间网络拓扑连通，通过 SR 的任意路径编排技术就可以实现对故障后通信路径的有效保护，这种技术与拓扑无关，因此称为拓扑无关无环备选（Topology Independent Loop-Free Alternate，TI-LFA）路径技术。

下面通过调整网络链路代价，构造邻居为 PQ 节点、远端邻居为 PQ 节点以及不存在PQ节点的情况，逐步开展LFA路径、R-LFA路径和TI-LFA路径技术实验。

8.2.1 实验环境

LFA 路径技术拓扑如图 8.2 所示，所有节点已按第 2 章的实验约定配置了接口 IP 地址、环回接口地址、MPLS 标签空间等，所有节点启用了 OSPF 协议并且均在区域 0，双向链路代价默认为 1。初始时未启动 LDP 和 OSPF-SR。

实验过程中将以节点 1 和节点 6 之间的链路为对象，在节点 1 上启动 FRR 相关功能，查看节点 1 上建立去往目的节点 6 的前缀 6.6.6.6/32 的保护路径。通过改变网络上个别链路的链路代价，影响各节点路由计算，进而观察在不同条件下，保护路径的变化情况，从而理解不同条件下的快速重路由技术。

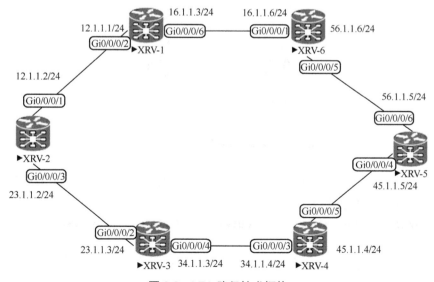

图 8.2　LFA 路径技术拓扑

8.2.2 LFA 路径技术

初始条件下,所有链路代价为1,现在调整Link(1,6)之间的双向链路代价为4,这种拓扑场景下,节点1针对目的前缀6.6.6.6/32的PQ节点集为{2, 3, 4, 5},如图 8.3 所示。将节点 1 作为 PLR(运行 LFA 路径计算的源节点),邻居节点 2 去往目的地的最优路径不经过节点 1,因此,当 Link(1,6)故障时,节点 1 只需要将数据分组直接转发给节点 2 就可以实现快速重路由。

PLR 计算 LFA 路径的规则为:Dis (N, D) < Dis (N, S) + Dis (S, D),其中,N 为 PLR 邻居节点,D 为目的节点,S 为运行 LFA 路径计算的源节点。Dis(X, Y) 表示从节点 X 到节点 Y 最短路径的距离,例如,Dis (N, D)表示从邻居节点 N 到目的节点 D 最短路径的距离。以上规则的意思为:从邻居节点 N 到目的节点 D 的距离比从邻居节点 N 到源节点 S 再加上 S 到目的节点 D 的距离短,即从邻居节点 N 到目的节点 D 的最短路径不会经过源节点 S。如果邻居节点满足上述规则,则该邻居满足链路保护条件,即流量不会从 LFA 节点重新回到源节点 S。本例中,Dis (N, D) = Dis(2,6) = 5,Dis (N, S) = Dis(2,1) = 1,Dis (S, D) = Dis(1,6) = 5,Dis(2,6)<Dis(2,1)+Dis(1, 6),所以,节点 2 满足作为 LFA 节点的条件。

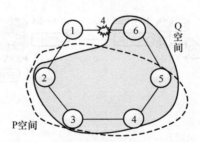

图 8.3 Cost(1,6)=4 时的 PQ 节点集

默认条件下,OSPF 没有启动快速重路由,此时可以查看节点 1 OSPF 的备用路由、路由表以及转发表:

```
RP/0/0/CPU0:XRV-1#show ospf routes 6.6.6.6/32 backup-path
O    6.6.6.6/32, metric 5
        16.1.1.6, from 6.6.6.6, via GigabitEthernet0/0/0/6, path-id 1
RP/0/0/CPU0:XRV-1#show route 6.6.6.6/32
```

```
Routing entry for 6.6.6.6/32
  Known via "ospf 1", distance 110, metric 5, type intra area
  Routing Descriptor Blocks
    16.1.1.6, from 6.6.6.6, via GigabitEthernet0/0/0/6
      Route metric is 5
  No advertising protos.
RP/0/0/CPU0:XRV-1#show cef 6.6.6.6/32
6.6.6.6/32, version 230, internal 0x1000001 0x0 (ptr 0xa13f1f54)
 [1], 0x0 (0xa13d77a0), 0x0 (0x0)
 local adjacency 16.1.1.6
 Prefix Len 32, traffic index 0, precedence n/a, priority 1
   via 16.1.1.6/32, GigabitEthernet0/0/0/6, 7 dependencies,
weight 0, class 0 [flags 0x0]
    path-idx 0 NHID 0x0 [0xa166e1c8 0x0]
    next hop 16.1.1.6/32
    local adjacency
```

可以通过如下指令开启 IP FRR 功能：

```
router ospf NAME fast-reroute per-prefix
```

在节点 1 上配置上述指令后，查看 OSPF 计算的 6.6.6.6/32 保护路径情况：

```
RP/0/0/CPU0:XRV-1#show ospf routes 6.6.6.6/32 backup-path
Topology Table for ospf 1 with ID 1.1.1.1

Codes: O - Intra area, O IA - Inter area
       O E1 - External type 1, O E2 - External type 2
       O N1 - NSSA external type 1, O N2 - NSSA external type 2

O    6.6.6.6/32, metric 5
       16.1.1.6, from 6.6.6.6, via GigabitEthernet0/0/0/6, path-id 1
   Backup path:
       12.1.1.2, from 6.6.6.6, via GigabitEthernet0/0/0/2, protected
bitmap 0000000000000001   Attributes: Metric: 6, SRLG Disjoint
```

可以看到 OSPF 进程已经计算得到保护路径，保护路径的下一跳为节点 2，出接口为 Gi0/0/0/2，保护路径代价为 6。进一步查看节点 1 的路由表和转发表信息：

243

```
RP/0/0/CPU0:XRV-1#show route 6.6.6.6/32


Routing entry for 6.6.6.6/32
  Known via "ospf 1", distance 110, metric 5, type intra area
  Installed Mar  2 17:29:09.220 for 00:35:31
  Routing Descriptor Blocks
    12.1.1.2, from 6.6.6.6, via GigabitEthernet0/0/0/2, Backup
(Local-LFA)
      Route metric is 6
    16.1.1.6, from 6.6.6.6, via GigabitEthernet0/0/0/6, Protected
      Route metric is 5
  No advertising protos.
RP/0/0/CPU0:XRV-1#do show cef 6.6.6.6/32
6.6.6.6/32, version 292, internal 0x1000001 0x0 (ptr 0xa13f1f54)
[1], 0x0 (0xa13d77a0), 0x0 (0x0)
 local adjacency 16.1.1.6
 Prefix Len 32, traffic index 0, precedence n/a, priority 1
   via 12.1.1.2/32, GigabitEthernet0/0/0/2, 6 dependencies,
weight 0, class 0, backup (Local-LFA) [flags 0x300]
    path-idx 0 NHID 0x0 [0xa166e16c 0x0]
    next hop 12.1.1.2/32
    local adjacency
   via 16.1.1.6/32, GigabitEthernet0/0/0/6, 6 dependencies,
weight 0, class 0, protected [flags 0x400]
    path-idx 1 bkup-idx 0 NHID 0x0 [0xa16da1fc 0x0]
    next hop 16.1.1.6/32
```

可以看到当前最优路由被设置了"Protected"标志，相应的保护路径标记为"Backup (Local-LFA)"，表示通过本地 LFA 路径算法计算得到。

8.2.3　Remote LFA（R-LFA）路径技术

现在将 Link(1,6) 的双向链路代价调整为 3 进行 R-LFA 实验。与上一种情况相比，在故障发生时，节点 1 无法通过 LFA 路径技术，将数据转发给邻居节点 2 而实现快速重路由，因为此时节点 2 去往节点 6 的最优路径会经过故障链路（节点 2→1→6）。查看节点 1 的 OSPF 进程，发现已经无法计算出保护路径，如下：

```
RP/0/0/CPU0:XRV-1#show ospf routes 6.6.6.6/32 backup-path

O    6.6.6.6/32, metric 5
       16.1.1.6, from 6.6.6.6, via GigabitEthernet0/0/0/6, path-id 1
```

需要启动远端 LFA 功能，以支持 FRR。

启动 R-LFA 的指令如下：

```
router ospf NAME fast-reroute per-prefix
router ospf NAME fast-reroute per-prefix remote-lfa tunnel mpls-ldp
```

需要先配置 fast-reroute，再启动 remote-lfa，否则可能会导致 R-LFA 不能正常工作。在节点 1 的 OSPF 进程开启 R-LFA 后，节点 1 针对目的前缀 6.6.6.6/32 的 PQ 节点集为{3, 4}，如图 8.4 所示。

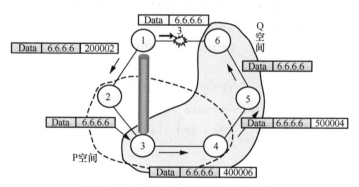

图 8.4　Cost(1,6)=3 时的 PQ 节点集

查看协议、路由和转发表信息：

```
RP/0/0/CPU0:XRV-1#show ospf routes 6.6.6.6/32 backup-path
O    6.6.6.6/32, metric 4
       16.1.1.6, from 6.6.6.6, via GigabitEthernet0/0/0/6, path-id 1
       Backup path: Remote, LFA: 3.3.3.3
       12.1.1.2, from 6.6.6.6, via GigabitEthernet0/0/0/2, protected
bitmap 0000000000000001  Attributes: Metric: 2, SRLG Disjoint
RP/0/0/CPU0:XRV-1#show route 6.6.6.6/32
```

```
Routing entry for 6.6.6.6/32
  Known via "ospf 1", distance 110, metric 4, type intra area
  Routing Descriptor Blocks
    12.1.1.2, from 6.6.6.6, via GigabitEthernet0/0/0/2, Backup
(Remote-LFA)
      Repair Node(s): 3.3.3.3
      Route metric is 2
    16.1.1.6, from 6.6.6.6, via GigabitEthernet0/0/0/6, Protected
      Route metric is 4
RP/0/0/CPU0:XRV-1#show cef 6.6.6.6/32
6.6.6.6/32, version 303, internal 0x1000001 0x0 (ptr 0xa13f1f54)
[1], 0x0 (0xa13d77a0), 0x0 (0x0)
 local adjacency 16.1.1.6
 Prefix Len 32, traffic index 0, precedence n/a, priority 1
   via 12.1.1.2/32, GigabitEthernet0/0/0/2, 9 dependencies,
weight 0, class 0, backup (Remote-LFA) [flags 0x8300]
    path-idx 0 NHID 0x0 [0xa166e16c 0x0]
    next hop 12.1.1.2/32, Repair Node(s): 3.3.3.3
    local adjacency
   via 16.1.1.6/32, GigabitEthernet0/0/0/6, 9 dependencies,
weight 0, class 0, protect-ignore [flags 0x400]
    path-idx 1 bkup-idx 0 NHID 0x0 [0xa166e1c8 0x0]
    next hop 16.1.1.6/32
    local adjacency
```

从 OSPF 进程和路由表信息来看，节点 1 已经计算出 R-LFA 路径，重定向流量选择在节点 3 释放。但是，从转发面的表项来看，主用路径的保护状态为"protect-ignore"，这说明保护路径虽然存在但是无法生效。根本的原因就是节点 1 缺少合适的数据面功能，将重定向流量引导至修复节点 3。实际上，从 R-LFA 路径的配置指令就可以看出，该功能是依赖于 MPLS 和 LDP 技术的。

R-LFA 路径技术的基本原理是基于 MPLS 的隧道能力，将流量引导至修复节点。一旦重路由的数据到达修复节点，就能沿其最优路径将数据送达目的地，且不会经过故障链路。

第 8 章 基于 SR 的 FRR 技术原理与实战

首先，在所有节点的所有接口上启动 LDP。LDP 启动后，使能 LDP 的接口便自动开启 MPLS 能力。此时，再次查看节点 1 的 MPLS 转发表项：

```
RP/0/0/CPU0:XRV-1#show mpls forwarding prefix 6.6.6.6/32 detail
Local   Outgoing     Prefix         Outgoing         Next Hop     Bytes
Label   Label        or ID          Interface                     Switched
------  -----------  -------------  -------------    ----------   --------
100004  Pop          6.6.6.6/32     Gi0/0/0/6        16.1.1.6     628
    Path Flags: 0x400 [ BKUP-IDX:0 (0xa16da288) ]
    Version: 275, Priority: 3
    Label Stack (Top -> Bottom): { Imp-Null }
    NHID: 0x0, Encap-ID: N/A, Path idx: 1, Backup path idx: 0,
Weight: 0
    MAC/Encaps: 14/14, MTU: 1500
    Outgoing Interface: GigabitEthernet0/0/0/6 (ifhandle 0x00000100)
    Packets Switched: 13

200002               6.6.6.6/32     Gi0/0/0/2        12.1.1.2     0 (!)
    Updated: Mar  2 19:06:48.449
    Path Flags: 0x300 [ IDX:0 BKUP, NoFwd ]
    Version: 275, Priority: 3
    Label Stack (Top -> Bottom): { 200002 Unlabelled }
    NHID: 0x0, Encap-ID: N/A, Path idx: 0, Backup path idx: 0,
Weight: 0
    MAC/Encaps: 14/18, MTU: 1500
    Outgoing Interface: GigabitEthernet0/0/0/2 (ifhandle 0x00000080)
    Packets Switched: 0

    (!): FRR pure backup
RP/0/0/CPU0:XRV-1#show cef 6.6.6.6/32
6.6.6.6/32, version 275, internal 0x1000001 0x0 (ptr 0xa13f1f54)
 [1], 0x0 (0xa13d77a0), 0xa28 (0xa170e368)
 local adjacency 16.1.1.6
 Prefix Len 32, traffic index 0, precedence n/a, priority 3
```

```
      via 12.1.1.2/32, GigabitEthernet0/0/0/2, 6 dependencies,
weight 0, class 0, backup [flags 0x300]
        path-idx 0 NHID 0x0 [0xa166e394 0x0]
        next hop 12.1.1.2/32
        local adjacency
        local label 100004         labels imposed {200002 None}
      via 16.1.1.6/32, GigabitEthernet0/0/0/6, 6 dependencies,
weight 0, class 0, protected [flags 0x400]
        path-idx 1 bkup-idx 0 NHID 0x0 [0xa16da288 0xa16da3a0]
        next hop 16.1.1.6/32
        local label 100004         labels imposed {ImplNull}
```

可以看到,基于 MPLS 的保护路径已经产生,且在数据转发面已经生效。

在节点 1 追踪标签 200002 信息,如下:

`RP/0/0/CPU0:XRV-1#show mpls ldp bindings remote-label 200002`

```
3.3.3.3/32, rev 231
      Local binding: label: 100001
      Remote bindings: (2 peers)
          Peer                    Label
          ----------------        --------
          2.2.2.2:0               200002
          6.6.6.6:0               500006
```

发现标签是绑定在前缀 3.3.3.3/32 上的,这个地址正是节点 3 的环回接口地址。

按照 OSPF 协议 R-LFA 路径的计算结果,节点 3 作为 6.6.6.6/32 的修复节点。追踪节点 2 分配的 200002 标签对应的 LSP 上相关节点的 MPLS 转发表以及修复节点 3 之后的 MPLS 转发表信息,如下:

```
RP/0/0/CPU0:XRV-2#show mpls forwarding labels 200002
Local    Outgoing    Prefix          Outgoing       Next Hop      Bytes
Label    Label       or ID           Interface                    Switched
------   --------    -----------     -----------    ----------    --------
200002   Pop         3.3.3.3/32      Gi0/0/0/3      23.1.1.3      180248

RP/0/0/CPU0:XRV-3#show mpls forwarding prefix 6.6.6.6/32
```

Local Label	Outgoing Label	Prefix or ID	Outgoing Interface	Next Hop	Bytes Switched
300008	400006	6.6.6.6/32	Gi0/0/0/4	34.1.1.4	0

RP/0/0/CPU0:XRV-4#show mpls forwarding labels 400006

Local Label	Outgoing Label	Prefix or ID	Outgoing Interface	Next Hop	Bytes Switched
400006	500004	6.6.6.6/32	Gi0/0/0/5	45.1.1.5	71978

RP/0/0/CPU0:XRV-5#show mpls forwarding labels 500004

Local Label	Outgoing Label	Prefix or ID	Outgoing Interface	Next Hop	Bytes Switched
500004	Pop	6.6.6.6/32	Gi0/0/0/6	56.1.1.6	10740

当故障发生时，节点 1 先将流量封装在以节点 3 为目的地的 MPLS 隧道中，将流量引导至节点 3；然后节点 3 将流量转发到最终目的地。保护路径的流量转发过程如图 8.4 所示，重定向的流量先压入 200002 标签后发给节点 2，节点 2 弹出标签还原为 IP 分组后转发给节点 3，节点 3 匹配目的 IP 地址 6.6.6.6/32，压入 400006 标签后转发给节点 4，节点 4 将标签交换为 500004 后转发节点 5，节点 5 弹出标签并还原为 IP 分组后转发给节点 6。

值得注意的是，保护路径标签栈自顶向下为{200002, Unlabelled}。这说明在使用保护路径进行数据转发时，不能简单地在原有分组头部压入保护路径标签，而应当将分组头部原有的所有标签全部剥离后，再压入保护路径标签。

通过下面的例子可以说明其作用。如图 8.5（a）所示，节点 1 收到目的 IP 地址为 5.5.5.5 且头部标签栈为{100004, 600002}的 MPLS 分组，正常条件下，节点 1 将头部标签 100004 剥离后转发给节点 6，节点 6 将 600002 剥离后将数据送至目的节点 5。当链路 Link(1,6)故障，使用保护路径转发分组时，假定节点 1 没有将原始分组的所有 MPLS 头剥离，而直接压入保护路径标签 200002，那么该分组的头部标签在节点 2 弹出发给节点 3 时，节点 3 会因为自己并没有分配该标签而丢弃该分组，或者因为该标签具有其他含义，而将分组引入其他未知路径上。总之，保护路径并未起到应有的保护作用。因此，节点 1 在进行快速重路由操作压入保护路径标签时，必须将原来所有的标签剥离掉，如图 8.5（b）所示，节点 1 将所有 MPLS 标签弹出后压入 200002，并沿着保护路径将数据分组送到目的节点 5，这就是保护路径标签栈栈底标签为"Unlabelled"的作用。

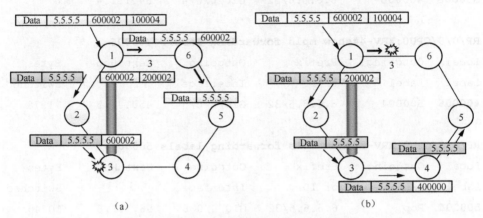

图 8.5 保护路径标签栈中 Unlabelled 作用

8.2.4 Extended R-LFA（ER-LFA）路径技术

在 R-LFA 路径基础上，进一步将 Link(1,6)的双向链路代价修改为 1，并将 Link(3,4)的双向链路代价修改为 2，按照 R-LFA 路径计算得到 P 空间和 Q 空间，如图 8.6（a）所示，此时 PQ 空间不相交，因此 PQ 节点不存在。但是，由于节点 2 存在到达 Q 空间中节点 4 的 LSP，且该 LSP 不经过 Link(1,6)，只要节点 1 将数据分组送入该 LSP 的入口，使得被保护的数据在节点 4 释放，也能正确将保护数据送达目的节点 6，如图 8.6（b）所示，这就是拓展 R-LFA（Extended R-LFA，ER-LFA）路径技术。

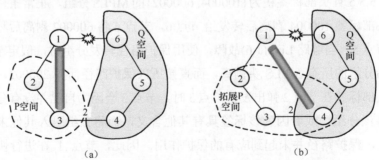

图 8.6 基本 PQ 空间与拓展 PQ 空间（节点 6 为目的地）

第 8 章 基于 SR 的 FRR 技术原理与实战

改变链路代价后，查看节点 1 的协议、路由和转发表信息：

```
RP/0/0/CPU0:XRV-1#show ospf routes 6.6.6.6/32 backup-path

O    6.6.6.6/32, metric 4
        16.1.1.6, from 6.6.6.6, via GigabitEthernet0/0/0/6, path-id 1
     Backup path: Remote, LFA: 4.4.4.4
        12.1.1.2, from 6.6.6.6, via GigabitEthernet0/0/0/2, protected
bitmap 0000000000000001 Attributes: Metric: 4, SRLG Disjoint
RP/0/0/CPU0:XRV-1#show route 6.6.6.6/32
Routing entry for 6.6.6.6/32
  Known via "ospf 1", distance 110, metric 2, type intra area
  Routing Descriptor Blocks
    12.1.1.2, from 6.6.6.6, via GigabitEthernet0/0/0/2, Backup
(Remote-LFA)
      Repair Node(s): 4.4.4.4
      Route metric is 3
    16.1.1.6, from 6.6.6.6, via GigabitEthernet0/0/0/6, Protected
      Route metric is 2
RP/0/0/CPU0:XRV-1#show mpls forwarding prefix 6.6.6.6/32 detail
Local   Outgoing    Prefix          Outgoing       Next Hop       Bytes
Label   Label       or ID           Interface                     Switched
------  ----------- --------------  -------------  -------------  --------
100004  Pop         6.6.6.6/32      Gi0/0/0/6      16.1.1.6       8410
    Path Flags: 0x400 [ BKUP-IDX:0 (0xa176b544) ]
    Version: 355, Priority: 3
    Label Stack (Top -> Bottom): { Imp-Null }
    NHID: 0x0, Encap-ID: N/A, Path idx: 1, Backup path idx: 0,
Weight: 0
    MAC/Encaps: 14/14, MTU: 1500
    Outgoing Interface: GigabitEthernet0/0/0/6 (ifhandle 0x00000100)
    Packets Switched: 172

200004  6.6.6.6/32       Gi0/0/0/2       12.1.1.2       0       (!)
    Updated: Mar  2 20:51:48.027
```

```
        Path Flags: 0x300 [ IDX:0 BKUP, NoFwd ]
        Version: 355, Priority: 3
        Label Stack (Top -> Bottom): { 200004 Unlabelled }
        NHID: 0x0, Encap-ID: N/A, Path idx: 0, Backup path idx: 0,
Weight: 0
        MAC/Encaps: 14/18, MTU: 1500
        Outgoing Interface: GigabitEthernet0/0/0/2 (ifhandle 0x00000080)
        Packets Switched: 0
        (!): FRR pure backup
```

可以看到，这个时候节点 1 针对 6.6.6.6/32 前缀计算的保护路径发生了变化，与 R-LFA 路径相比，其将修复节点修改为节点 4，对应保护路径标签栈由 {200002 Unlabelled} 改为 {200004 Unlabelled}。在节点 1 追踪标签 200004 信息，如下：

```
RP/0/0/CPU0:XRV-1#show mpls ldp bindings remote-label 200004


4.4.4.4/32, rev 232
        Local binding: label: 100002
        Remote bindings: (2 peers)
          Peer                      Label
          -----------------         --------
          2.2.2.2:0                 200004
          6.6.6.6:0                 600000
```

可以看到标签 200004 是绑定在前缀 4.4.4.4/32 上的，这个地址正是节点 4 的环回接口地址。追踪节点 2 分配的 200004 标签对应的 LSP 上相关节点的 MPLS 转发表以及修复节点 4 之后的 MPLS 转发表信息，如下：

```
RP/0/0/CPU0:XRV-2#show mpls forwarding labels 200004
Local   Outgoing    Prefix          Outgoing      Next Hop      Bytes
Label   Label       or ID           Interface                   Switched
------  ----------  --------------  ------------  ------------  --------
200004  300004      4.4.4.4/32      Gi0/0/0/3     23.1.1.3      3702


RP/0/0/CPU0:XRV-3#show mpls forwarding labels 300004
Local   Outgoing    Prefix          Outgoing      Next Hop      Bytes
```

第 8 章 基于 SR 的 FRR 技术原理与实战

```
Label    Label      or ID         Interface              Switched
300004   Pop        4.4.4.4/32    Gi0/0/0/4   34.1.1.4   51690

RP/0/0/CPU0:XRV-4#show mpls forwarding prefix 6.6.6.6/32
Local    Outgoing   Prefix        Outgoing    Next Hop   Bytes
Label    Label      or ID         Interface              Switched
400006   500004     6.6.6.6/32    Gi0/0/0/5   45.1.1.5   71978

RP/0/0/CPU0:XRV-5#show mpls forwarding labels 500004
Local    Outgoing   Prefix        Outgoing    Next Hop   Bytes
Label    Label      or ID         Interface              Switched
500004   Pop        6.6.6.6/32    Gi0/0/0/6   56.1.1.6   10740
```

基于查询信息，将标签操作标注于保护路径上，如图 8.7 所示，当故障发生时，节点 1 将重定向的流量压入 200004 标签后发给节点 2，节点 2 交换为出标签 300004 后转发给节点 3，节点 3 弹出标签还原为 IP 分组后转发给节点 4，节点 4 匹配目的 IP 地址 6.6.6.6/32，压入 500004 标签后转发给节点 5，节点 5 弹出标签并还原为 IP 分组后转发给节点 6，进而实现对相应流量的保护。

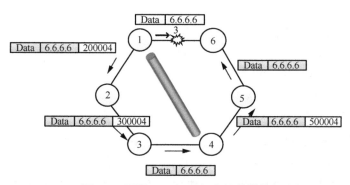

图 8.7 拓展 R-LFA 路径上标签操作

8.2.5 Topology Independent LFA（TI-LFA）路径技术

在 ER-LFA 路径技术基础上，进一步将 Link(3,4) 的双向链路代价修改为 4，其他链路代价为 1，按照 ER-LFA 路径计算得到 P 空间和 Q 空间，如图 8.8 所示，此时 PQ 节点不存在。

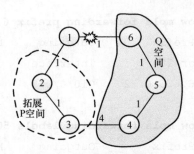

图 8.8　ER-LFA 失效场景

查看节点 1 OSPF 进程的保护路径计算情况，发现已经无法计算出保护路径。

```
RP/0/0/CPU0:XRV-1#show ospf routes 6.6.6.6/32 backup-path
O    6.6.6.6/32, metric 5
     16.1.1.6, from 6.6.6.6, via GigabitEthernet0/0/0/6, path-id 1
```

需要启动 SR 功能，基于 SR 技术实现 TI-LFA 路径，进而实现与拓扑无关的保护路径计算。

配置 TI-LFA 路径的相关指令如下：

```
router ospf NAME segment-routing mpls
router ospf NAME segment-routing forwarding mpls
router ospf NAME fast-reroute per-prefix
router ospf NAME fast-reroute per-prefix ti-lfa [enable|disable]
```

在各节点配置启动 SR，并在节点开启 TI-LFA 路径功能，查看节点 1 上 OSPF 进程计算的保护路径信息和保护路径标签栈信息，如下：

```
RP/0/0/CPU0:XRV-1#show ospf routes 6.6.6.6/32 backup-path
O    6.6.6.6/32, metric 2
     16.1.1.6, from 6.6.6.6, via GigabitEthernet0/0/0/6, path-id 1
     Backup path: TI-LFA, Repair-List: P node: 3.3.3.3   Label: 16003
                                       P node: 4.4.4.4   Label: 16004
         12.1.1.2, from 6.6.6.6, via GigabitEthernet0/0/0/2, protected
bitmap 0000000000000001 Attributes: Metric: 9, SRLG Disjoint
```

第 8 章 基于 SR 的 FRR 技术原理与实战

```
RP/0/0/CPU0:XRV-1#show mpls forwarding prefix 6.6.6.6/32 detail
Local      Outgoing     Prefix         Outgoing       Next Hop        Bytes
Label      Label        or ID          Interface                      Switched
------     --------     ---------      -----------    ----------      --------
100004     Pop          6.6.6.6/32     Gi0/0/0/6      16.1.1.6        0
     Path Flags: 0x40400 [ PROT-IGNORE ]
     Version: 563, Priority: 15
     Label Stack (Top -> Bottom): { Imp-Null }
     NHID: 0x0, Encap-ID: N/A, Path idx: 1, Backup path idx: 0,
Weight: 0
     MAC/Encaps: 14/14, MTU: 1500
     Outgoing Interface: GigabitEthernet0/0/0/6 (ifhandle 0x00000100)
     Packets Switched: 0
     Protection is ignored

200002                  6.6.6.6/32     Gi0/0/0/2      12.1.1.2        0 (!)
     Path Flags: 0x300 [ IDX:0 BKUP, NoFwd ]
     Version: 563, Priority: 15
     Label Stack (Top -> Bottom): { 200002 16004 16006 }
     NHID: 0x0, Encap-ID: N/A, Path idx: 0, Backup path idx: 0,
Weight: 0
     MAC/Encaps: 14/26, MTU: 1500
     Outgoing Interface: GigabitEthernet0/0/0/2 (ifhandle 0x00000080)
     Packets Switched: 0
     (!): FRR pure backup
```

可以看到 OSPF 基于 TI-LFA 路径技术又可以计算出保护路径了。该保护路径计算得到的 P 节点为 3，通过 200002 标签将流量引导至节点 3，然后通过 16004 标签引导流量至节点 4，再通过 16006 标签将保护流量引导至原始下一跳节点 6。从标签栈操作来说，节点 1 在原有分组前压入 3 层标签{200002, 16004, 16006}，栈顶标签 200002 为 LDP 分配标签，用于引导流量到达节点 3；标签 16004 为 SR 标签，指导节点 3 将流量引导至节点 4；16006 标签用于将流量引导至原始下一跳节点 6，TI-LFA 路径上的标签操作如图 8.9 所示。

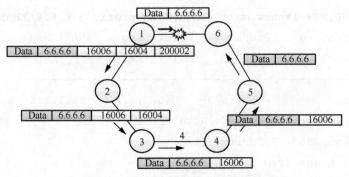

图 8.9 TI-LFA 路径上的标签操作

这里，在生成保护路径标签栈时，采用了 SR 标签和 LDP 标签混合方式，200002 标签为 LDP 分配，用于引导流量到达节点 3，可以通过指令禁用这种混合标签模式，而仅使用 SR-OSPF 分配的标签来构建保护标签栈。

设置偏好 SR 标签的指令为：

```
router ospf NAME segment-routing sr-prefer
```

在节点 1 使能偏好 SR 标签后，查看保护路径标签栈信息：

```
RP/0/0/CPU0:XRV-1#show mpls forwarding prefix 6.6.6.6/32 detail
Local    Outgoing    Prefix          Outgoing      Next Hop       Bytes
Label    Label       or ID           Interface                    Switched
------   ---------   -------------   -----------   -----------    -----
100004   Pop         6.6.6.6/32      Gi0/0/0/6     16.1.1.6       0
     Path Flags: 0x400 [ BKUP-IDX:0 (0xa176b42c) ]
     Version: 608, Priority: 1
     Label Stack (Top -> Bottom): { Imp-Null }
     NHID: 0x0, Encap-ID: N/A, Path idx: 1, Backup path idx: 0,
Weight: 0
     MAC/Encaps: 14/14, MTU: 1500
     Outgoing Interface: GigabitEthernet0/0/0/6 (ifhandle 0x00000100)
     Packets Switched: 0

16003            SR Pfx (idx 6)     Gi0/0/0/2     12.1.1.2       0(!)
     Path Flags: 0xb00 [ IDX:0 BKUP, NoFwd ]
     Version: 608, Priority: 1
```

第 8 章 基于 SR 的 FRR 技术原理与实战

```
    Label Stack (Top -> Bottom): { 16003 16004 16006 }
    NHID: 0x0, Encap-ID: N/A, Path idx: 0, Backup path idx: 0,
Weight: 0
    MAC/Encaps: 14/26, MTU: 1500
    Outgoing Interface: GigabitEthernet0/0/0/2 (ifhandle 0x00000080)
    Packets Switched: 0
    (!): FRR pure backup
```

可以看到标签栈信息由原来的{200002, 16004, 16006}更改为{16003, 16004, 16006}，全部采用 SR-OSPF 分配的标签。

下面在节点 1 上通过 Nil-FEC 进行路径追踪，查看保护路径的情况：

```
RP/0/0/CPU0:XRV-1#traceroute mpls nil-fec labels 16003,16004,16006
output interface gigabitEthernet 0/0/0/2 nexthop 12.1.1.2

Tracing MPLS Label Switched Path with Nil FEC with labels [16003,
16004,16006], timeout is 2 seconds

  0 12.1.1.1 MRU 1500 [Labels: 16003/16004/16006/explicit-null Exp:
0/0/0/0]
L 1 12.1.1.2 MRU 1500 [Labels: implicit-null/16004/16006/explicit-
null Exp: 0/0/0/0] 0 ms
L 2 23.1.1.3 MRU 1500 [Labels: implicit-null/16006/explicit-null
Exp: 0/0/0] 0 ms
L 3 34.1.1.4 MRU 1500 [Labels: 16006/explicit-null Exp: 0/0] 10 ms
L 4 45.1.1.5 MRU 1500 [Labels: implicit-null/explicit-null Exp:
0/0] 10 ms
! 5 56.1.1.6 10 ms
```

该路径上各节点的 MPLS 交换表信息为，节点 1 发出携带标签栈为{16003, 16004, 16006, explicit-null}的 MPLS 分组；节点 2 将栈顶标签弹出后，发出标签栈为{16004, 16006, explicit-null}的分组；节点 3 将栈顶标签弹出后，发出标签栈为{16006, explicit-null}的分组；节点 4 将栈顶标签交换为 16006 后，发出标签栈为{16006, explicit-null}的分组；节点 5 弹出栈顶标签后，发出标签栈为 explicit-null 的分组；节点 6 弹出显式空标签后，匹配目的 IP 地址 127.0.0.1 到本地协议栈 OAM 进程进行处理，并回应源节点。可以看到保护路径经过的节点列表为节点 1→2→3→4→5→6，各节点的转发标签与预期一致。

与 R-LFA 路径技术相比，TI-LFA 的保护路径标签栈里没有"Unlabelled"标签，这是因为 TI-LFA 路径技术计算得到的保护路径标签栈栈底标签为被保护前缀对应的标签值，它可以确保到被重定向的数据分组能够保持原样送达原始下一跳节点，并由原始节点对保护分组进行正确的解析和后续处理。保护路径栈底标签的作用如图 8.10 所示，对照图 8.5 的示例，节点 1 进行保护路径数据传送时，保留了原始分组中携带的标签信息 600002。通过保护路径将数据送达节点 6 后，再由节点 6 对标签 600002 进行解释和转发。如果单纯从数据转发路径来看，R-LFA 路径为节点 1→2→3→4→5，而 TI-LFA 路径为节点 1→2→3→4→5→6→5。比较两者发现，TI-LFA 路径使得数据转发绕路了。实际上，当存在多层标签时，一些标签并非仅用于指导数据转发，而是与一些业务特性关联，例如，节点 6 是一个包检测设备，它匹配标签 600002 后要对该分组进行深度检测后再转发。当采用 R-LFA 路径技术时，节点 1 为了对数据进行保护传输，剥离了该标签值，使得数据未经过节点 6 的检测就送达目的节点 5；而采用 TI-LFA 路径技术时，仍然可以保证数据先经过节点 6 的检测再送达目的地。两者相比，R-LFA 路径只保证数据能够送达最终目的节点，而 TI-LFA 路径技术在实现对数据保护传输的同时，可以最大化保留原始数据分组中预置的信息，因而具备更好的特性。

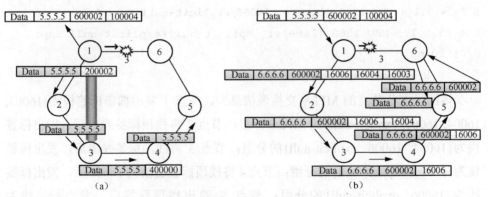

图 8.10 保护路径栈底标签的作用

第 8 章 基于 SR 的 FRR 技术原理与实战

针对图 8.10（b）场景，在节点 1 上通过 Nil-FEC 模拟生成一个被保护的数据分组，其携带保护路径标签和原始标签，标签栈为{16003，16004，16006，600002}，通过路径追踪，查看保护路径的情况：

```
RP/0/0/CPU0:XRV-1#traceroute mpls nil-fec labels 16003,16004,16006,
600002 output interface gigabitEthernet 0/0/0/2 nexthop 12.1.1.2

Tracing MPLS Label Switched Path with Nil FEC with labels [16003,
16004,16006,600002], timeout is 2 seconds

  0 12.1.1.1 MRU 1500 [Labels: 16003/16004/16006/600002/explicit-
null Exp: 0/0/0/0/0]
L 1 12.1.1.2 MRU 1500 [Labels: implicit-null/16004/16006/600002/
explicit-null Exp: 0/0/0/0/0] 10 ms
L 2 23.1.1.3 MRU 1500 [Labels: implicit-null/16006/600002/explicit-
null Exp: 0/0/0/0] 0 ms
L 3 34.1.1.4 MRU 1500 [Labels: 16006/600002/explicit-null Exp:
0/0/0] 10 ms
L 4 45.1.1.5 MRU 1500 [Labels: implicit-null/600002/explicit-null
Exp: 0/0/0] 10 ms
L 5 56.1.1.6 MRU 1500 [Labels: implicit-null/explicit-null Exp:
0/0] 10 ms
! 6 56.1.1.5 10 ms
```

可以看到保护数据分组按照预期的保护路径，先经过节点 5，然后携带 600002 标签到达节点 6，由节点 6 处理完后，再送给节点 5，其处理过程与不经过保护路径时的预期处理过程一致。

为了进一步实验 AdjSID 在 TI-LFA 路径中的作用，下面修改 Link(3,4)的双向链路代价，将其设置为 6，此时节点 3 如果使用节点 4 的 NodeSID 引导流量，会将流量引导经过 Link(1,6)，而无法实现路径保护。这个时候 AdjSID 的作用就体现出来了。

查看节点 1 上 OSPF 进程计算的保护路径信息和保护路径标签栈信息：

```
RP/0/0/CPU0:XRV-1#show ospf routes 6.6.6.6/32 backup-path
O    6.6.6.6/32, metric 2
```

```
           16.1.1.6, from 6.6.6.6, via GigabitEthernet0/0/0/6, path-id 1
      Backup path: TI-LFA, Repair-List: P node: 3.3.3.3   Label: 16003
                                        Q node: 4.4.4.4   Label: 300012
           12.1.1.2, from 6.6.6.6, via GigabitEthernet0/0/0/2, protect
ed bitmap 0000000000000001 Attributes: Metric: 11, SRLG Disjoint
RP/0/0/CPU0:XRV-1#show mpls forwarding prefix 6.6.6.6/32 detail
Local   Outgoing  Prefix            Outgoing       Next Hop      Bytes
Label   Label     or ID             Interface                    Switched
------  --------  ----------------  -------------  ------------  --------
16006   Pop       SR Pfx (idx 6)    Gi0/0/0/6      16.1.1.6      0
     Path Flags: 0x400 [ BKUP-IDX:0 (0xa176b42c) ]
     Version: 620, Priority: 1
     Label Stack (Top -> Bottom): { Imp-Null }
     NHID: 0x0, Encap-ID: N/A, Path idx: 1, Backup path idx: 0,
Weight: 0
     MAC/Encaps: 14/14, MTU: 1500
     Outgoing Interface: GigabitEthernet0/0/0/6 (ifhandle 0x00000100)
     Packets Switched: 0

16003             SR Pfx (idx 6)    Gi0/0/0/2      12.1.1.2      0(!)
     Path Flags: 0xb00 [ IDX:0 BKUP, NoFwd ]
     Version: 620, Priority: 1
     Label Stack (Top -> Bottom): { 16003 300012 16006 }
     NHID: 0x0, Encap-ID: N/A, Path idx: 0, Backup path idx: 0,
Weight: 0
     MAC/Encaps: 14/26, MTU: 1500
     Outgoing Interface: GigabitEthernet0/0/0/2 (ifhandle 0x00000080)
     Packets Switched: 0
     (!): FRR pure backup
```

可以看到保护路径发生了变化，该保护路径计算得到的 P 节点为节点 3，Q 节点为节点 4，通过 16003 标签将流量引导至节点 3，然后通过 300012 将流量引导至节点 4，最后通过 16006 将保护流量引导至原始下一跳（节点 6）。因为标签 300012 为节点 3 引导流量至节点 4 的标签，这个标签必然是节点 3 分配

的标签,可以在节点 1 上查看节点 3 分配的标签与链路的映射关系:

```
RP/0/0/CPU0:XRV-1#show ospf database opaque-area adv-router 3.3.3.3
            OSPF Router with ID (1.1.1.1) (Process ID 1)
             Type-10 Opaque Link Area Link States (Area 0)
LS age: 438
Options: (No TOS-capability, DC)
LS Type: Opaque Area Link
Link State ID: 8.0.0.7
Opaque Type: 8
Opaque ID: 7
Advertising Router: 3.3.3.3
LS Seq Number: 80000007
Checksum: 0xdbeb
Length: 76
  Extended Link TLV: Length: 52
    Link-type : 2                      //Transit Network
    Link ID   : 34.1.1.3
    Link Data : 34.1.1.3
  LAN Adj sub-TLV: Length: 11
      Flags       : 0xe0  //为到非 DR 邻居分配具有保护能力的标签
      MTID        : 0
      Weight      : 0
      Neighbor ID : 4.4.4.4
      Label       : 300013
  LAN Adj sub-TLV: Length: 11
      Flags       : 0x60  //为到非 DR 邻居分配不具有保护能力的标签
      MTID        : 0
      Weight      : 0
      Neighbor ID : 4.4.4.4
      Label       : 300012
  Link MSD sub-TLV: Length: 2
    Type: 1, Value 10
```

从节点 1 的 OSPF 链路状态数据库中可以看到,节点 3 针对与节点 4 的链路分配两个标签，一个是具有保护能力的标签 300013，另一个为不具有保护能力的标签 300012（当本地启动 TI-LFA 时才会分配具有保护能力的标签）。节点 1 在生成保护路径标签栈时使用了不具有保护能力的标签。TI-LFA 路径上标签操作如图 8.11 所示，从标签栈操作来说，节点 1 在原有分组前压入 3 层标签{16003, 300012, 16006}，栈顶标签 16003 用于引导流量到达节点 3，标签 300012 用于指导节点 3 将流量引导至邻居节点 4，16006 标签用于将流量引导至原始下一跳节点 6。

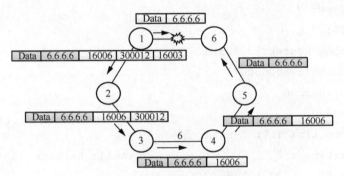

图 8.11　TI-LFA 路径上标签操作

8.2.6　TI-LFA 路径的数据面验证

FRR 技术通过本地算法，基于网络拓扑数据，在本地计算生成两条路由：最优路由和保护路由，或称为主用路由和备用路由。这两条路由将以主备模式全部下发到转发面。通常情况下，采用主用路由进行数据转发，一旦路由失效，则立刻启用备用路由。

主备路由切换的条件是主用路由失效。那么发生何种事件时判定为主用路由失效？通常的判定依据就是与主用路由关联的接口进入 Down 状态，或者将主用路由跟 BFD 关联，在接口 Up 条件下依据某些业务的可达性触发主备路由切换。

下面以图 8.11 为基础，构建 FRR 的数据面验证拓扑如图 8.12 所示，进行 TI-LFA 场景下的快速重路由实验，注意 Link(3,4)的双向链路代价仍为 6。首

第 8 章　基于 SR 的 FRR 技术原理与实战

先，在节点 1 上连接一台虚拟个人计算机（Virtual Personal Computer，VPC），由该主机向目的节点 6 进行 IP ping 操作。然后，在节点 1 上关闭 Gi0/0/0/6 接口模拟 Link(1,6)故障，触发节点 1 进行主备路由切换。

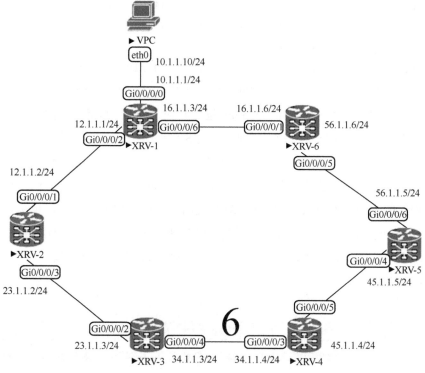

图 8.12　FRR 的数据面验证拓扑

在节点 1 的 Gi0/0/0/6 接口启动快速接口关闭功能，以便在接口 Down 后触发主备路由切换。启动指令如下：
`interface IFNAME fast-shutdown`

在 VPC 上 ping 目的地址 6.6.6.6，设置 ping 包发送间隔为 50ms，在发送 ping 包的同时，关闭接口 Gi0/0/0/6，得到的响应结果如下：

```
VPCS> ping 6.6.6.6 -i 50 -t

84 bytes from 6.6.6.6 icmp_seq=1 ttl=254 time=12.864 ms
84 bytes from 6.6.6.6 icmp_seq=2 ttl=254 time=9.889 ms
```

```
84 bytes from 6.6.6.6 icmp_seq=3 ttl=254 time=4.029 ms
84 bytes from 6.6.6.6 icmp_seq=4 ttl=254 time=3.420 ms
…
84 bytes from 6.6.6.6 icmp_seq=151 ttl=254 time=3.671 ms
84 bytes from 6.6.6.6 icmp_seq=152 ttl=254 time=8.840 ms
84 bytes from 6.6.6.6 icmp_seq=153 ttl=254 time=5.740 ms
6.6.6.6 icmp_seq=154 timeout
84 bytes from 6.6.6.6 icmp_seq=155 ttl=250 time=53.496 ms
84 bytes from 6.6.6.6 icmp_seq=156 ttl=250 time=16.480 ms
```

可以看到，除了 seq=154 的 ping 包，其他 ping 包都得到了回应。与此同时，在节点 1 的 Gi0/0/0/2 接口上抓包，采用 SR 保护路径转发的数据分组如图 8.13 所示。可以看到，ICMP Request 序号为 154（包序号 747）的 ping 包被节点 1 正常转发出去，而没有丢失，只是没有得到节点 6 的响应（因为节点 6 尚未更新路由，相应的响应包通过其接口 Gi0/0/0/1 发出，进而丢失）。同时，也可以看到这个 ping 包被压入了备用路径标签栈{16003, 300012, 16006}，说明节点 1 已经快速从主用路径切换到备用路径上，并按照备用路径进行数据转发。

```
    Time         Source      Destination    Li Protocol   Length  Info
745 2898.6021... 12.1.1.1    224.0.0.5         OSPF       106    Hello Packet
746 2899.7919... 12.1.1.2    224.0.0.5         OSPF       106    Hello Packet
747 2901.2660... 10.1.1.10   6.6.6.6           ICMP       110    Echo (ping) request  id=0xcc07, seq=154/39424,
748 2901.3504... 12.1.1.1    224.0.0.5         OSPF       122    LS Update
749 2901.4001... 12.1.1.1    224.0.0.5         OSPF       130    LS Update
750 2902.3164... 10.1.1.10   6.6.6.6           ICMP       110    Echo (ping) request  id=0xcd07, seq=155/39680,
751 2902.3165... 12.1.1.2    224.0.0.5         OSPF       130    LS Update
752 2902.3600... 6.6.6.6     10.1.1.10         ICMP       98     Echo (ping) reply    id=0xcd07, seq=155/39680,
753 2902.4150... 10.1.1.10   6.6.6.6           ICMP       110    Echo (ping) request  id=0xcd07, seq=156/39936,
754 2902.4285... 6.6.6.6     10.1.1.10         ICMP       98     Echo (ping) reply    id=0xcd07, seq=156/39936,

Frame 747: 110 bytes on wire (880 bits), 110 bytes captured (880 bits) on interface -, id 0
Ethernet II, Src: 50:00:00:01:00:03 (50:00:00:01:00:03), Dst: 50:00:00:02:00:02 (50:00:00:02:00:02)
MultiProtocol Label Switching Header, Label: 16003, Exp: 0, S: 0, TTL: 63
MultiProtocol Label Switching Header, Label: 300012, Exp: 0, S: 0, TTL: 63
MultiProtocol Label Switching Header, Label: 16006, Exp: 0, S: 1, TTL: 63
Internet Protocol Version 4, Src: 10.1.1.10, Dst: 6.6.6.6
Internet Control Message Protocol
    Type: 8 (Echo (ping) request)
    Code: 0
    Checksum: 0x536a [correct]
    [Checksum Status: Good]
    Identifier (BE): 52231 (0xcc07)
    Identifier (LE): 1996 (0x07cc)
    Sequence Number (BE): 154 (0x009a)
    Sequence Number (LE): 39424 (0x9a00)
  > [No response seen]
  > Data (56 bytes)
```

图 8.13　采用 SR 保护路径转发的数据分组

由于实验拓扑网络规模很小，节点很快就能计算出收敛路径。在计算出收敛路径后，将删除主备路由，使用收敛后的路由进行数据转发。因此，很难捕获到采用 TI-LFA 路径进行数据转发的过程。上述过程是在延迟了节点 1 的 SPF 计算时间后模拟出来的。

OSPF SPF 延迟计算的配置指令如下：

```
router ospf NAME timers throttle spf D1 D2 D3
D1: <1-600000> Delay between receiving a change to SPF calculation in milliseconds
D2: <1-600000> Delay between first and second SPF calculation in milliseconds
D3: <1-600000> Maximum wait time in milliseconds for SPF calculations
```

在进行上述实验时，一定要先在节点 1 上配置 SPF 延迟计算才能捕获上述现象。

通过 TI-LFA 路径技术可以保证在出现链路故障后，节点在计算出新的收敛路径之前的这段时间内，到达故障上游节点的数据分组能够正确送达原始下一跳节点。当节点计算出最新的路由后，将撤销已下发的主备路由，并使用新路由进行数据转发。网络中不同节点采用新的收敛路由进行数据转发的时机不一样，于是就会产生收敛期间的微环路问题。

8.3 防微环路

在网络发生故障或故障恢复期间，IGP 都会重新收敛。由于采用分布式计算，受各个节点持有的网络拓扑数据不一致、最短路径算法计算快慢不一等因素制约，每个节点计算出收敛路径的时间是不一样的，在这个时间差内，局部节点对网络拓扑的认知出现矛盾，这一矛盾导致数据转发在局部节点之间出现微环路问题，使得网络节点之间转发状态出现短暂不一致，进而产生转发微环路现象，导致端到端通信数据丢失。

8.3.1 TI-LFA 路径与微环路

以图 8.12 拓扑为例，通过实验分析在节点 1 启动了 TI-LFA 路径后，当

Link(1,6)故障时，协议收敛期间微环路对通信的影响。为了模拟协议收敛期间节点之间路由不一致而产生的微环路问题，仅在节点 3 上配置了 SPF 延迟计算，设置 SPF 计算时延为 10s，也就是说，节点 3 在收到链路状态更新后 10s 才开始 SPF 计算，并更新路由。

实验过程中由 VPC 向目的节点 6 进行 IP ping 操作，然后，在节点 1 上关闭 Gi0/0/0/6 接口模拟 Link(1,6)故障，触发节点 1 进行主备路由切换和整个网络收敛。在节点 1 的 Gi0/0/0/2 接口和节点 6 的 Gi0/0/0/5 接口进行数据包捕获，通过对比数据包节点 1 发出的 ping 数据和节点 6 上收到的 ping 数据，验证环路对数据转发的影响。

VPC 上的 ping 结果如下，可以看到在 Link(1,6)关闭后，VPC 发出的 Seq=97～106 的 ICMP ping 都没有得到响应。

```
VPCS> ping 6.6.6.6 -i 50 -t
…
84 bytes from 6.6.6.6 icmp_seq=96 ttl=254 time=5.666 ms
6.6.6.6 icmp_seq=97 timeout
6.6.6.6 icmp_seq=98 timeout
6.6.6.6 icmp_seq=99 timeout
6.6.6.6 icmp_seq=100 timeout
6.6.6.6 icmp_seq=101 timeout
6.6.6.6 icmp_seq=102 timeout
6.6.6.6 icmp_seq=103 timeout
6.6.6.6 icmp_seq=104 timeout
6.6.6.6 icmp_seq=105 timeout
6.6.6.6 icmp_seq=106 timeout
84 bytes from 6.6.6.6 icmp_seq=107 ttl=250 time=48.856 ms
```

节点 1 发出与节点 6 收到的分组如图 8.14 所示，左侧为节点 1 发出的数据分组，可以看到 Seq=97～106 的 ping 包都从节点 1 发出。从右侧节点 6 收到的数据分组来看，Seq=97 的分组被收到后，下一个就是 Seq=107 的分组。这说明两个问题：一是节点 1 通过快速重路由机制，能够保证当前节点不会因为故障而丢弃收到的分组，所以 Seq=97 的分组没有丢失；二是协议收敛期间，因为收敛时间不一致，导致节点 1～节点 6 之间出现了微环路，造成 Seq=98～106 这些分组丢失。

第 8 章 基于 SR 的 FRR 技术原理与实战

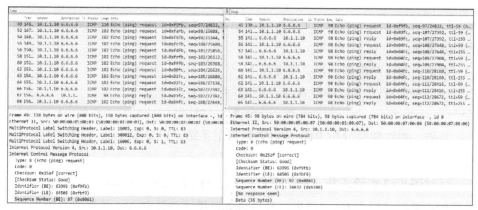

图 8.14　节点 1 发出与节点 6 收到的分组

进一步探究 Seq=97 的分组封装，Seq=97 和 98 分组的标签封装如图 8.15 所示。在图 8.15 左侧，发现节点 1 是通过保护路径标签栈{16003, 300012, 16006}来引导该分组绕过故障链路，进而将其送达节点 6。而对于 Seq=98～106 的这些分组，节点 1 采用了收敛路径进行转发，见图 8.15 右侧，即使用{16006}标签来引导流量，由于收敛路径受微环路影响，也无法将分组送到目的地。

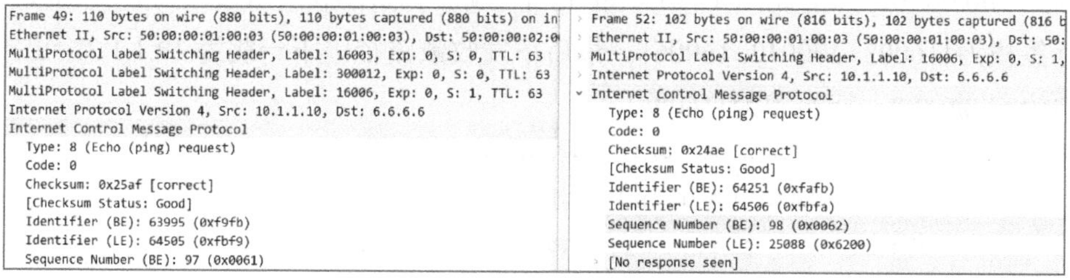

图 8.15　Seq=97 和 98 分组的标签封装

8.3.2　TI-LFA 路径与本地防微环路

通过上面的实验得到的一个直接启示是，既然 TI-LFA 路径技术能够保证数据持续送到目的节点，那么就让节点一直使用该保护路径进行数据转发，等到整个网络收敛了再使用收敛路径进行数据转发，这样就可以避免过早弃用保护路径使用收敛路径带来的丢包问题。这个方法被称为本地防微环路机制。

267

RFC8333 给出了该机制的阐述，即路由信息库（Routing Information Base，RIB）的延迟更新。

本地防微环路机制的相关指令如下：

```
router ospf NAME microloop avoidance
router ospf NAME microloop avoidance rib-update-delay <1-600000>
```

首先，通过实验分析 RIB 延迟更新与主备路径以及 SPF 计算的收敛路径的关系。

针对图 8.12 实验拓扑，在节点 1 上启动 TI-LFA 路径和本地防微环路机制，配置 RIB 更新时延为 10s。首先，开启 Debug 模式，让 Debug 信息输出至串口，然后关闭 Link(1,6)接口，接口关闭时间为 22:17:31.068。接下来启动 SPF 路径计算，首先设置下一次 SPF 计算的时延为 50ms，然后在 22:17:31.158 调度 SPF 算法，并在 22:17:31.228 开始 Dijkstra 计算。由于网络规模很小，立刻计算出收敛路由。但是收敛路由并没有立刻下发，而是等到 22:17:41.257 才开始 RIB 更新，恰恰延迟了 10s，这就是 RIB 延迟更新避免微环路的作用。下面再看 RIB 的更新过程。RIB 更新时，首先删除了原来下发的主用路径，下一跳为 16.1.1.6，出接口为 Gi0/0/0/6；然后删除 TI-LFA 路径，下一跳为 12.1.1.2，出接口为 Gi0/0/0/2，标签栈为{16003, 300012, 16006}。最后下发收敛路径，下一跳为 12.1.1.2，出接口为 Gi0/0/0/2，标签栈为{16006}。

```
RP/0/0/CPU0:XRV-1#debug ospf 1 monitor
RP/0/0/CPU0:XRV-1#debug rib routing
RP/0/0/CPU0:XRV-1(config)#logging console debugging
RP/0/0/CPU0:XRV-1(config)#inter g0/0/0/6
RP/0/0/CPU0:XRV-1(config-if)#shu
RP/0/0/CPU0:XRV-1(config-if)#commit
RP/0/0/CPU0:Mar 14 22:17:31.068 : ifmgr[228]: %PKT_INFRA-LINK-5-
CHANGED : Interface GigabitEthernet0/0/0/6, changed state to
Administratively Down
RP/0/0/CPU0:XRV-1(config-if)#RP/0/0/CPU0:Mar 14 22:17:31.068 :
ospf[1018]:   reset throttling to 50 ms
......
```

```
RP/0/0/CPU0:Mar 14 22:17:31.158 : ospf[1018]: sched dijkstra:
Schedule SPF in area 0 Change in LS ID 1.1.1.1, LSA type R,
RP/0/0/CPU0:Mar 14 22:17:31.158 : ospf[1018]:  reset throttling to
50 ms
RP/0/0/CPU0:Mar 14 22:17:31.158 : ospf[1018]:  Schedule SPF: area
0spf_time 000047533.519856454 wait_interval 000000000.050000000
RP/0/0/CPU0:Mar 14 22:17:31.228 : ospf[1018]:  Begin Dijkstra for
area 0 at 000047626.313499654
RP/0/0/CPU0:Mar 14 22:17:31.228 : ospf[1018]:  End Dijkstra at
000047626.313499654, Total elapsed time 000000000.000000000
RP/0/0/CPU0:Mar 14 22:17:36.087 : ospf[1018]:  reset throttling to
50 ms
......
RP/0/0/CPU0:Mar 14 22:17:41.257 : ipv4_rib[1148]: RIB Routing: Vrf:
"default", Tbl: "default" IPv4 Unicast, Delete active route to
6.6.6.6 via 16.1.1.6 interface GigabitEthernet0/0/0/6, metric
[110/2], label 3, by client-id 15
RP/0/0/CPU0:Mar 14 22:17:41.277 : ipv4_rib[1148]: RIB Routing: Vrf:
"default", Tbl: "default" IPv4 Unicast, Delete active route to
6.6.6.6 via 12.1.1.2 interface GigabitEthernet0/0/0/2, metric
[110/11], label 16003 300012 16006, by client-id 15
RP/0/0/CPU0:Mar 14 22:17:41.277 : ipv4_rib[1148]: RIB Routing: Vrf:
"default", Tbl: "default" IPv4 Unicast, Add active route 6.6.6.6/32
via 12.1.1.2 interface GigabitEthernet0/0/0/2, metric [110/11]
(fl: 0xd800000/0x40000) label 16006, by client ospf
```

下面，实验验证节点 1 在启动了 TI-LFA 路径和本地防微环路机制后的端到端数据传送情况。

从 VPC 的 ping 来看，在 Link(1,6)关闭后，只有 Seq=59 的分组没有得到响应。

```
VPCS> ping 6.6.6.6 -i 50 -t
…
84 bytes from 6.6.6.6 icmp_seq=57 ttl=254 time=8.505 ms
84 bytes from 6.6.6.6 icmp_seq=58 ttl=254 time=4.255 ms
```

```
6.6.6.6 icmp_seq=59 timeout
84 bytes from 6.6.6.6 icmp_seq=60 ttl=250 time=23.023 ms
```

节点 1 发出和节点 6 收到的分组如图 8.16 所示。可以看到 Seq=59 的分组被节点 1 发出，且到达了节点 6，只是节点 6 针对该分组的响应还是通过 Link(6,1) 发出，而且在链路上丢失了。节点 1 持续使用 TI-LFA 路径进行数据转发，并在网络收敛之后才启用收敛路径进行数据转发，使得从链路失效到网络重新收敛的整个时间段内，从节点 1 发向节点 6 的数据分组一个都没有丢失。这充分说明了通过 TI-LFA 路径和本地防微环路，可以有效解决链路故障引起的路由黑洞和协议收敛引起的路由环路问题。

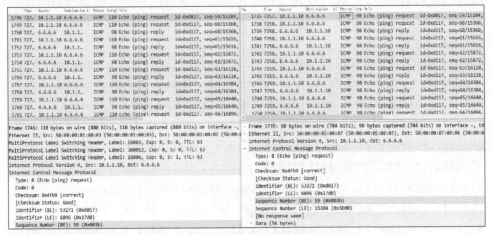

图 8.16　节点 1 发出和节点 6 收到的分组

8.3.3　TI-LFA 路径与 SR 防微环路

对于故障链路的邻接节点而言，它能够通过预先计算并下发主备路径，结合本地防微环路的技术，消除整个故障事件对端到端通信的影响。但是对于故障链路的非邻接节点，它无法快速感知远端故障，也不可能针对网络中所有的拓扑变化预先计算保护路径，因此仍然无法保证拓扑变化期间端到端数据传送的可靠性。

当然，链路故障只是网络拓扑变化的一种条件，链路恢复、节点故障或者恢复等都会引发网络拓扑变化，这些拓扑变化都会引起网络中所有节点重新进行路由计算和协议收敛，因此，都会产生微环路问题。下面就以链路故障和恢复期间引发的微环路为实验场景，阐述网络拓扑变化过程的微环路以及基于 SR 的防微环路效果。

1．链路故障时 SR 防微环路

针对图 8.12 实验拓扑，在节点 2 上也启动本地防微环路功能，RIB 更新时延与节点 1 一致，都是 10s。节点 2 向节点 6 发送 ping 包，在节点 1 上关闭 Gi0/0/0/6 接口，查看节点 2 到节点 6 之间数据的可达性，可以看到 Seq=487～491 之间的 ping 没有得到响应，如下：

```
RP/0/0/CPU0:XRV-2#ping 6.6.6.6 repeat 1000000 verbose
…
Reply to request 484 (1 ms)
Reply to request 485 (1 ms)
Reply to request 486 (1 ms)
Request 487 timed out
Request 488 timed out
Request 489 timed out
Request 490 timed out
Request 491 timed out
Reply to request 492 (79 ms)
Reply to request 493 (9 ms)
```

节点 1 在 Gi0/0/0/2 接口上收到的分组如图 8.17 所示，可以看到节点 1 收到了包序号 8058 的分组（Seq=487），它是节点 2 发向节点 1 的分组，携带 16006 标签；包序号 8059 的分组为节点 1 向节点 2 转发的 8058 号分组，携带 {16003,300012,16006} 标签，说明这个时候节点 1 检测到 Link(1,6) 故障，并启动了 TI-LFA 路径，于是，节点 1 收到的携带 MPLS 标签为 16006 的分组被交换成 16006 标签后，又压入了保护路径标签 300012 和 16003，这 3 个标签将引导该分组避开故障，将数据送达目的节点。8058 号分组是节点 1 从节点 2 收到的最后一个分组，说明之后节点 2 更新了本地 RIB，并按照新的路径进行数据转发。

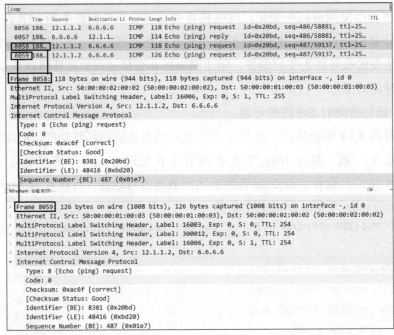

图 8.17 节点 1 在 Gi0/0/0/2 接口上收到的分组

节点 6 在 Gi0/0/0/5 接口上收到的分组如图 8.18 所示，它正常接收到上述 Seq=487 的分组，说明节点 2 发给节点 1 的 ICMP Request 分组经过保护路径的保护，正常到达节点 6。同时，节点 6 收到的下一个 ICMP Request 分组 Seq=492，说明节点 2 在使用更新后的 RIB 进行数据转发时在节点 2 与节点 3，甚至其他节点之间遭遇了环路，从而导致这些分组丢失。

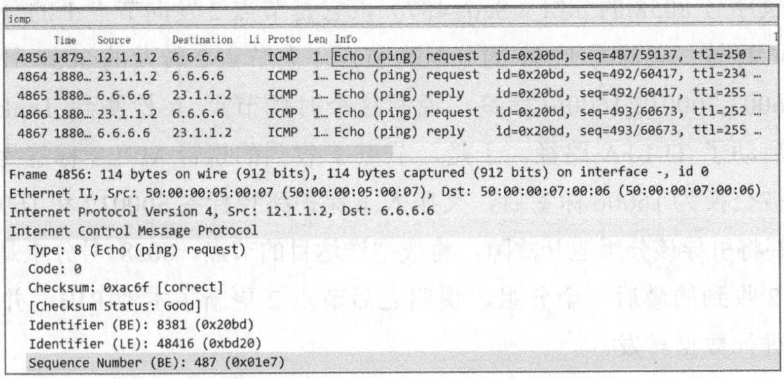

图 8.18 节点 6 在 Gi0/0/0/5 接口上收到的分组

这一现象充分说明了本地防微环路无法彻底解决微环路问题。分析一下上述过程，节点 2 在更新 RIB 前，持续使用原始路径将分组发给节点 1，节点 1 使用保护路径传送，数据持续可达。在节点 1 和节点 2 的 RIB 更新时延结束之后，节点 1 和节点 2 分别进行 RIB 更新，都使用收敛后的路径进行数据转发。这里面一个关键的点就是节点 1 和节点 2 使用收敛路径的时机，因为节点 1 和节点 2 无法实现严格时钟同步，而且两者的计算路由耗时也不同，虽然 RIB 时延是一样的，但由于计算路由的耗时不一样，进而导致节点 1 和节点 2 的收敛路由更新时间不会严格一致。如果节点 1 先使用了收敛路径，而节点 2 仍然使用收敛前的路径，那么就会在节点 1 和节点 2 之间产生环路。当然，节点 1 启用收敛路径的前提是其他节点都完成了收敛，如果按照这个逻辑反推，节点 2 应该先于节点 1 完成 RIB 更新。既然节点 2 先于节点 1 完成 RIB 更新，那么节点 2 就会将一部分流量发送给节点 3，避免这部分流量在节点 2 和节点 3 之间产生环路的条件就是节点 3 先于节点 2 完成 RIB 更新，节点 3 要先于节点 4 完成 RIB 更新，依次类推。只有这样才能保证节点 2 切换到收敛路径上时不会出现环路。这实际上是非常苛刻的，完全依靠本地 RIB 更新时延无法实现这种排序性的 RIB 更新。

要剥离掉其他节点依次 RIB 更新的苛刻要求，需要有新的思路。方法很简单，回顾一下节点 1 使用 TI-LFA 路径+本地防微环路机制能够保证持续可达的原因，就是计算出的收敛路径不要立刻使用，而是等待。那么对于节点 2 而言，是否也可以采用相同的思路，即不要立刻使用收敛路径来转发数据，而是采用类似于 TI-LFA 路径来转发数据，让数据转发对 RIB 更新次序不产生依赖性。这种路径称为 SR 防微环路，当收到拓扑变化消息但未计算出 SR 防微环路和收敛路径时，先采用原始路径进行数据转发，此时的数据会到达 PLR 节点，通过保护路径确保数据可达；然后下发并使用 SR 防微环路，通过 SR 防微环路保证数据可达；最后，在 SR 防微环路到期后，再撤销 SR 防微环路，并下发收敛路径，使用收敛路径进行数据转发，从而保证路径的持续可达性。

SR 防微环路的相关指令如下：

```
ipv4 unnumbered mpls traffic-eng Loopback0
mpls traffic-eng
router ospf NAME microloop avoidance segment-routing
router ospf NAME microloop avoidance rib-update-delay <1-600000>
```

SR 防微环路以流量工程（Traffic Engineering，TE）隧道路由的形式来使用，因此，必须要在节点上启动 TE 功能。

在节点 2 配置 SR 防微环路功能，打开 Debug 模式，关闭节点 1 的 Gi0/0/0/6 接口后记录的 RIB 更新信息如下：

```
RP/0/0/CPU0:Mar  4 09:35:00.001 : ospf[1018]: Begin Dijkstra for
area 0 at 000156766.199107988
RP/0/0/CPU0:Mar  4 09:35:00.011 : ospf[1018]: End Dijkstra at
000156766.209107303, Total elapsed time 000000000.009999315
......
RP/0/0/CPU0:Mar  4 09:35:00.991 : ipv4_rib[1148]: RIB Routing: Vrf:
"default", Tbl: "default" IPv4 Unicast, Delete active route to
6.6.6.6 via 12.1.1.1 interface GigabitEthernet0/0/0/1, metric
[110/3], label 16006, by client-id 15
RP/0/0/CPU0:Mar  4 09:35:00.991 : ipv4_rib[1148]: RIB Routing: Vrf:
"default", Tbl: "default" IPv4 Unicast, Add active route 6.6.6.6/32
via 0.0.0.0 interface tunnel-te32839, metric [110/10] (fl: 0xd800000/
0x40000) label 16006, by client ospf
RP/0/0/CPU0:Mar  4 09:35:00.991 : ipv4_rib[1148]: RIB Routing: Vrf:
"default", Tbl: "default" IPv4 Unicast, Update local-label 16006
(1) to 6.6.6.6/32 by proto ospf client ospf
......
RP/0/0/CPU0:Mar  4 09:35:10.020 : ipv4_rib[1148]: RIB Routing: Vrf:
"default", Tbl: "default" IPv4 Unicast, Delete active route to
6.6.6.6 via 0.0.0.0 interface tunnel-te32839, metric [110/10],
label 16006, by client-id 15
RP/0/0/CPU0:Mar  4 09:35:10.020 : ipv4_rib[1148]: RIB Routing: Vrf:
"default", Tbl: "default" IPv4 Unicast, Add active route 6.6.6.6/32
via 23.1.1.3 interface GigabitEthernet0/0/0/3, metric [110/10] (fl:
0xd800000/0x40000) label 16006, by client ospf
RP/0/0/CPU0:Mar  4 09:35:10.020 : ipv4_rib[1148]: RIB Routing: Vrf:
"default", Tbl: "default" IPv4 Unicast, Update local-label 16006
(1) to 6.6.6.6/32 by proto ospf client ospf
```

从上述过程中可以看到，节点 2 在收到网络拓扑变化事件后，启动 SPF 算

第 8 章 基于 SR 的 FRR 技术原理与实战

法,计算完毕后立刻(09:35:00.991)删除去往 6.6.6.6/32 的路由(下一跳 12.1.1.1),同时添加一条隧道 tunnel-te32839,这条隧道就是 SR 防微环路。SR 防微环路持续使用 10s 后,在 09:35:10.020 时刻,删除该隧道,并下发收敛路由(下一跳 23.1.1.3)。

当节点 2 在使用 SR 防微环路进行数据转发时,此时查看 OSPF 备用路由计算结果、路由表和转发表信息如下:

```
RP/0/0/CPU0:XRV-2#show ospf routes 6.6.6.6/32 backup-path
Topology Table for ospf 1 with ID 2.2.2.2
O    6.6.6.6/32, metric 10
        23.1.1.3, from 6.6.6.6, via GigabitEthernet0/0/0/3, path-id 1
            SR microloop path: Repair-List: P node: 3.3.3.3   Label: 3
                                            Q node: 4.4.4.4   Label: 300012
                                            microloop Tunnel: tunnel-te32840

RP/0/0/CPU0:XRV-2#show ip route
O     1.1.1.1/32 [110/2] via 12.1.1.1, 12:27:55, GigabitEthernet0/0/0/1
L     2.2.2.2/32 is directly connected, 1d19h, Loopback0
O     3.3.3.3/32 [110/2] via 23.1.1.3, 12:27:55, GigabitEthernet0/0/0/3
O     4.4.4.4/32 [110/8] via 0.0.0.0, 00:00:08, tunnel-te32840
O     5.5.5.5/32 [110/9] via 0.0.0.0, 00:00:08, tunnel-te32840
O     6.6.6.6/32 [110/10] via 0.0.0.0, 00:00:08, tunnel-te32840

RP/0/0/CPU0:XRV-2#show mpls forwarding prefix 6.6.6.6/32 detail
Thu Mar  4 10:14:07.220 UTC
Local  Outgoing    Prefix           Outgoing    Next Hop        Bytes
Label  Label       or ID            Interface                   Switched
------ ----------- ---------------- ----------- --------------- -----------
16006  16006       SR Pfx (idx 6)   tt32840     point2point     0
       Version: 4610, Priority: 1
       Label Stack (Top -> Bottom): { Unlabelled 16006 }
       NHID: 0x0, Encap-ID: N/A, Path idx: 0, Backup path idx: 0,
Weight: 0
       MAC/Encaps: 0/4, MTU: 0
       Outgoing Interface: tunnel-te32840 (ifhandle 0x000009b0)
```

```
    Packets Switched: 0

  Traffic-Matrix Packets/Bytes Switched: 0/0
```

可以看到，此时 OSPF 计算得到去往 6.6.6.6/32 的备用路径为 SR 防微环路，且本地去往节点 4～节点 6 的路由均使用 SR 防微环路，该路径被标识为 tunnel-te32840。去往 6.6.6.6/32 的 MPLS 标签操作为：弹出所有标签后压入 16006 标签，然后将分组送入 tunnel-te32840。此时 tunnel-te32840 信息如下：

```
RP/0/0/CPU0:XRV-2#show mpls traffic-eng tunnels
Name: tunnel-te32840  Destination: 4.4.4.4  Ifhandle:0x970 (auto-
tunnel for OSPF 1)
  Signalled-Name: auto_XRV-2_t32840
  Status:
    Admin:    up Oper:    up  Path: valid    Signalling: connected
    path option 10, (verbatim Segment-Routing) type explicit
(_te32840) (Basis for Setup)
    G-PID: 0x0800 (derived from egress interface properties)
    Bandwidth Requested: 0 kbps  CT0
    Creation Time: Thu Mar  4 09:56:14 2021 (00:03:01 ago)
  Config Parameters:
    Bandwidth:        0 kbps (CT0) Priority:  7  7 Affinity: 0x0/
0xffff
    Metric Type: TE (global)
    Path Selection:
      Tiebreaker: Min-fill (default)
      Protection: any (default)
    Hop-limit: disabled
    Cost-limit: disabled
    Path-invalidation timeout: 10000 msec (default), Action: Tear
(default)
    AutoRoute: disabled  LockDown: disabled   Policy class: not set
    Forward class: 0 (default)
    Forwarding-Adjacency: disabled
    Autoroute Destinations: 0
```

第 8 章　基于 SR 的 FRR 技术原理与实战

```
      Loadshare:            0 equal loadshares
      Auto-bw: disabled
      Path Protection: Not Enabled
      BFD Fast Detection: Disabled
      Reoptimization after affinity failure: Enabled
      SRLG discovery: Disabled
   History:
      Tunnel has been up for: 00:03:01 (since Thu Mar 04 09:56:14
   UTC 2021)
      Current LSP:
         Uptime: 00:03:01 (since Thu Mar 04 09:56:14 UTC 2021)

   Segment-Routing Path Info (IGP information is not used)
      Segment0[First Hop]: 23.1.1.3, Label: -
      Segment1[ - ]: Label: 300012
   Displayed 1 (of 1) heads, 0 (of 0) midpoints, 0 (of 0) tails
```

可以看到该隧道为 auto-tunnel，由 OSPF 1 创建，隧道在 10000ms 后设置为无效，这与 RIB 更新时延是一致的。该隧道段列表为{3, 300012}，即送入该隧道的分组将被压入 300012 标签后转发给下一跳节点 23.1.1.3。

综合 MPLS 转发表信息和隧道信息，节点 2 将收到栈顶标签为 16006 的 MPLS 分组或者去往 6.6.6.6/32 的 IP 分组的操作总结为：如果携带标签，则将分组的所有标签剥离掉，然后压入{16006, 300012}，转发给下一跳节点 23.1.1.3。节点 3 捕获的节点 2 发向 SR 防微环路的分组如图 8.19 所示，节点 2 使用 SR 防微环路时，观察在节点 3 的 Gi0/0/0/2 接口捕获的分组，可以看到节点 2 的处理与上述描述一致。

2. 链路恢复时 SR 防微环路

链路恢复时 SR 防微环路实验拓扑如图 8.20 所示，除了 Link(5,6)链路权重为 6，其他链路权重均为 1。为了模拟链路恢复之后协议收敛期间节点之间路由不一致而产生的微环路现象，在节点 2 上配置 SPF 计算时延为 10s，其他节点的 SPF 延迟计算参数采用默认值。也就是说节点 2 在收到链路状态更新后 10s 才开始 SPF 计算，并更新路由，其他节点以默认方式处理，即在收到链路更新后立刻计算路由，并在得到路由后立刻使用新路由转发数据。

```
24 28.809600  23.1.1.2  6.6.6.6    ICMP    122 Echo (ping) request  id=0xc0bd, seq=62/15872,
25 28.820892  6.6.6.6   23.1.1.2   ICMP    114 Echo (ping) reply    id=0xc0bd, seq=62/15872,
26 28.826481  23.1.1.2  6.6.6.6    ICMP    122 Echo (ping) request  id=0xc0bd, seq=63/16128,

Frame 24: 122 bytes on wire (976 bits), 122 bytes captured (976 bits) on interface -, id 0
Ethernet II, Src: 50:00:00:02:00:04 (50:00:00:02:00:04), Dst: 50:00:00:03:00:03 (50:00:00:03:00:03)
MultiProtocol Label Switching Header, Label: 300012, Exp: 0, S: 0, TTL: 255
MultiProtocol Label Switching Header, Label: 16006,  Exp: 0, S: 1, TTL: 255
Internet Protocol Version 4, Src: 23.1.1.2, Dst: 6.6.6.6
Internet Control Message Protocol
    Type: 8 (Echo (ping) request)
    Code: 0
    Checksum: 0x0e18 [correct]
    [Checksum Status: Good]
    Identifier (BE): 49341 (0xc0bd)
    Identifier (LE): 48576 (0xbdc0)
    Sequence Number (BE): 62 (0x003e)
    Sequence Number (LE): 15872 (0x3e00)
```

图 8.19　节点 3 捕获的节点 2 发向 SR 防微环路的分组

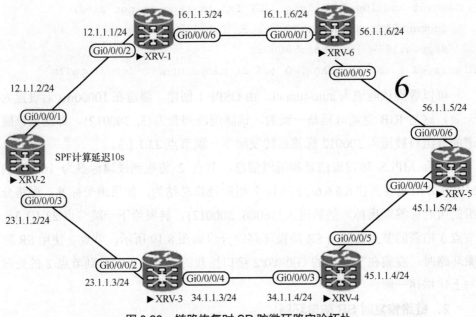

图 8.20　链路恢复时 SR 防微环路实验拓扑

为了更好地模拟链路故障，将节点 1 的 Gi0/0/0/6 和节点 6 的 Gi0/0/0/1 接口网络类型设置为点到点类型。指令如下：

```
router ospf NAME area X interface ifNUM network point-to-point
```

实验开始前，在节点 6 上关闭 Gi0/0/0/1 接口，模拟 Link(1,6)链路处于关闭状态，所有节点启动 SR，但未启动 SR-FRR 和 SR 防微环路，由节点 3 向目的

第 8 章 基于 SR 的 FRR 技术原理与实战

节点 6 进行 IP ping 操作。实验开始后，在节点 6 上使能 Gi0/0/0/1 接口，模拟 Link(1,6)恢复，触发整个网络收敛。在节点 2 的 Gi0/0/0/1 接口和节点 6 的 Gi0/0/0/5 接口进行抓包，通过对比节点 3 发出的 ping 数据和节点 6 上收到的 ping 数据，分析链路恢复时路由收敛期间的环路对数据转发的影响。

节点 6 与节点 3 捕获的数据分组如图 8.21 所示，左上为节点 6 的 Gi0/0/0/5 接口上抓到的包，可以看到，一开始节点 3 去往目的节点 6 的下一跳为 34.1.1.4，它使用源 IP 地址 34.1.1.3 作为 ICMP ping 的源地址，Seq=193 之前的 ping 分组都沿着节点 3→4→5→6 到达节点 6 的 Gi0/0/0/5 接口。图 8.21 右下为节点 2 的 Gi0/0/0/1 接口上抓到的包，可以看到，Link(1,6)链路恢复后，网络重新收敛，节点 3 将去往节点 6 的下一跳指向节点 2，所以 Seq=193 之后的 ping 分组都转发给节点 2。但是由于节点 2 采用了 SPF 延迟计算，导致其路由更新晚于节点 3 的路由更新，进而在节点 2 和节点 3 之间产生了路由环路，引发了丢包。所以，在环路消除后，节点 2 转发的第一个 ICMP ping 分组的 Seq=195，也就是 Seq=194 的分组因为环路而丢弃了。可见，只要在网络拓扑发生变化时，都有可能引发微环路，而不是只有在链路出现故障时才会引发。

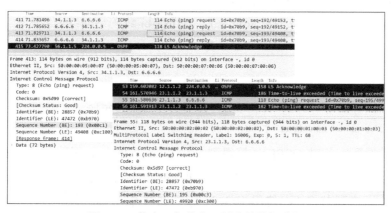

图 8.21 节点 6 与节点 3 捕获的数据分组

下面在节点 3 上启动 SR 防微环路功能，并重复上述实验操作，关闭节点 6 的 Gi0/0/0/1 接口后查看 OSPF 备用路由计算结果、路由表和转发表信息：

```
RP/0/0/CPU0:XRV-3#show ospf routes 6.6.6.6/32 backup-path
Topology Table for ospf 1 with ID 3.3.3.3
O  6.6.6.6/32, metric 4
```

```
          23.1.1.2, from 6.6.6.6, via GigabitEthernet0/0/0/2, path-id 1
      SR microloop path: Repair-List: P node: 1.1.1.1   Label: 16001
                                      Q node: 6.6.6.6   Label: 100001
                                      microloop Tunnel: tunnel-te32794
RP/0/0/CPU0:XRV-3#show route 6.6.6.6/32
Routing entry for 6.6.6.6/32
  Known via "ospf 1", distance 110, metric 4, labeled SR, type intra
area
  Routing Descriptor Blocks
    directly connected, via tunnel-te32794
      Route metric is 4
  No advertising protos.
RP/0/0/CPU0:XRV-3#show mpls forwarding prefix 6.6.6.6/32 detail
Local   Outgoing   Prefix              Outgoing         Next Hop        Bytes
Label   Label      or ID               Interface                        Switched
------  --------   ----------------    ------------     -------------   --------
16006   Pop        SR Pfx (idx 6)      tt32794          point2point     0
     Version: 4588, Priority: 1
     Label Stack (Top -> Bottom): { Unlabelled Imp-Null }
     NHID: 0x0, Encap-ID: N/A, Path idx: 0, Backup path idx: 0,
Weight: 0
     MAC/Encaps: 0/0, MTU: 0
     Outgoing Interface: tunnel-te32794 (ifhandle 0x00000370)
     Packets Switched: 0
```

可以看到，此时 OSPF 计算得到去往 6.6.6.6/32 的备用路径为 SR 防微环路径，该路径被标识为 tunnel-te32794，通过两个标签引导数据穿过该隧道：使用 16001 标签到节点 1，使用 100001 标签到节点 6。查看节点 1 的标签分配信息：

```
RP/0/0/CPU0:XRV-3#show ospf database opaque-area adv-router 1.1.1.1
            OSPF Router with ID (3.3.3.3) (Process ID 1)
                Type-10 Opaque Link Area Link States (Area 0)
......
  LS age: 225
```

```
Options: (No TOS-capability, DC)
LS Type: Opaque Area Link
Link State ID: 8.0.0.10
Opaque Type: 8
Opaque ID: 10
Advertising Router: 1.1.1.1
LS Seq Number: 80000001
Checksum: 0x8f12
Length: 80

  Extended Link TLV: Length: 56
    Link-type : 1            //network type: point-to-point
    Link ID     : 6.6.6.6
    Link Data : 16.1.1.1

    Adj sub-TLV: Length: 7
      Flags       : 0x60
      MTID        : 0
      Weight      : 0
      Label       : 100001

    Local-ID Remote-ID Private sub-TLV: Length: 12
      Local Interface ID: 10
      Remote Interface ID: 5

    Remote If Address sub-TLV: Length: 4
      Neighbor Address: 16.1.1.6

    Link MSD sub-TLV: Length: 2
      Type: 1, Value 10
```

节点 1 为到邻居 6.6.6.6 的链路分配 100001 标签，该链路的对端 IP 地址为 16.1.1.6，本地地址为 16.1.1.1。即从 OSPF 协议计算结果来看，期望先将数据分组通过 PrefixSID=16001 引导至节点 1，然后通过节点 1 的 AdjSID=100001 引导至节点6。

节点 3 去往 6.6.6.6/32 的 MPLS 标签操作为：弹出所有标签后分组送入 tunnel-te32794。此时 tunnel-te32794 信息如下：

```
RP/0/0/CPU0:XRV-3#show mpls traffic-eng tunnels
Name: tunnel-te32794  Destination: 6.6.6.6  Ifhandle:0x3b0 (auto-tu
nnel for OSPF 1)
  Signalled-Name: auto_XRV-3_t32794
  Status:
……
  Segment-Routing Path Info (IGP information is not used)
    Segment0[First Hop]: 23.1.1.2, Label: -
    Segment1[ - ]: Label: 16001
    Segment2[ - ]: Label: 100001
```

可以看到该隧道为 auto-tunnel，由 OSPF 1 创建，段列表为{16001, 100001}，即送入该隧道的分组将先后压入 100001 和 16001 标签，然后转发给下一跳节点 23.1.1.2。

综合节点 3 的 MPLS 转发表信息和隧道信息，可以将它收到栈顶标签为 16006 的 MPLS 分组或者去往 6.6.6.6/32 的 IP 分组的操作总结为：如果携带标签，则将分组的所有标签剥离掉，然后压入{16001, 100001}，转发给下一跳节点 23.1.1.2。节点 2 捕获的节点 3 发向 SR 微环路的分组如图 8.22 所示，可以看到节点 2 从节点 3 收到的第一个分组（Seq=41）采用了{16001, 100001}标签栈封装，但是 TTL 值为 179，说明节点 3 在使用 tunnel-te32794 隧道转发之前，按照原来的路由将 Seq=41 的分组交给了下一跳（节点 4），而节点 4 已经更新了路由，所以将该分组又送回了节点 3，于是在节点 3 和节点 4 之间出现了短暂的环路，但是这个时间非常短，并没有导致该分组 TTL 超时。当该分组再次环回到节点 3 时，节点 3 采用了隧道转发，使用{16001, 100001}标签栈将该分组引导至目的节点而没有丢弃。当节点 3 发出 Seq=42 的分组时，TTL 值为 255，短暂环路不再发生，节点 3 发出的 ping 分组能够使用 SR 防微环路，确保数据持续送到节点 6，直到 RIB 更新定时器到期后，再采用收敛路径进行数据转发。

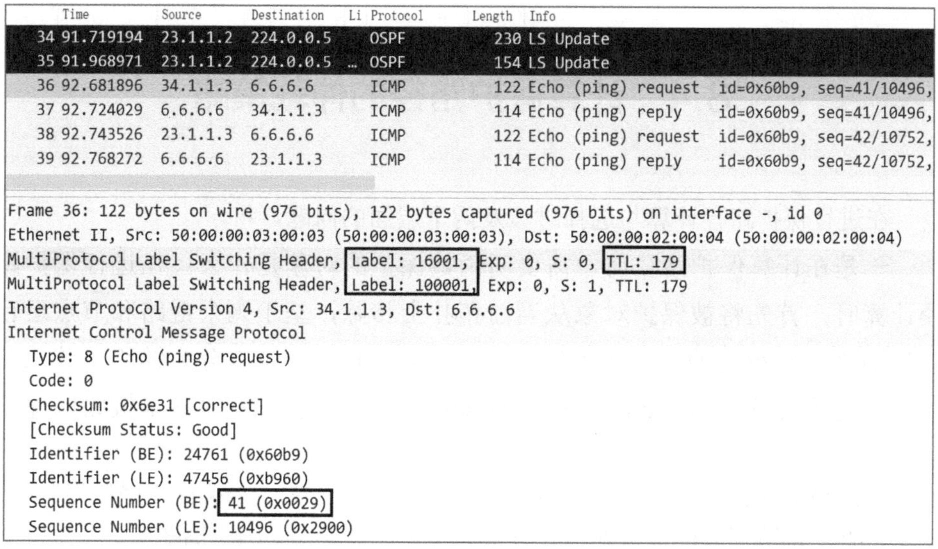

第 8 章 基于 SR 的 FRR 技术原理与实战

```
   Time         Source      Destination  Li Protocol  Length Info
34 91.719194   23.1.1.2    224.0.0.5       OSPF      230 LS Update
35 91.968971   23.1.1.2    224.0.0.5    …  OSPF      154 LS Update
36 92.681896   34.1.1.3    6.6.6.6         ICMP      122 Echo (ping) request  id=0x60b9, seq=41/10496,
37 92.724029   6.6.6.6     34.1.1.3        ICMP      114 Echo (ping) reply    id=0x60b9, seq=41/10496,
38 92.743526   23.1.1.3    6.6.6.6         ICMP      122 Echo (ping) request  id=0x60b9, seq=42/10752,
39 92.768272   6.6.6.6     23.1.1.3        ICMP      114 Echo (ping) reply    id=0x60b9, seq=42/10752,
```

```
Frame 36: 122 bytes on wire (976 bits), 122 bytes captured (976 bits) on interface -, id 0
Ethernet II, Src: 50:00:00:03:00:03 (50:00:00:03:00:03), Dst: 50:00:00:02:00:04 (50:00:00:02:00:04)
MultiProtocol Label Switching Header, Label: 16001, Exp: 0, S: 0, TTL: 179
MultiProtocol Label Switching Header, Label: 100001, Exp: 0, S: 1, TTL: 179
Internet Protocol Version 4, Src: 34.1.1.3, Dst: 6.6.6.6
Internet Control Message Protocol
    Type: 8 (Echo (ping) request)
    Code: 0
    Checksum: 0x6e31 [correct]
    [Checksum Status: Good]
    Identifier (BE): 24761 (0x60b9)
    Identifier (LE): 47456 (0xb960)
    Sequence Number (BE): 41 (0x0029)
    Sequence Number (LE): 10496 (0x2900)
```

图 8.22　节点 2 捕获的节点 3 发向 SR 防微环路的分组

防微环路与 TI-LFA 路径技术类似，均通过基于 SR 的任意流量引导功能实现对数据转发的保护。但是两者又存在不同。TI-LFA 路径技术是特定节点针对特定保护对象预先计算保护路径，并在保护对象失效时，使用保护路径进行快速重路由；防微环路功能则可以在任何一个节点开启，且不需要预先设定保护对象。通常情况下，当网络中的节点收到网络拓扑更新消息后，会调用 SPF 算法计算得到新的收敛路径，并将最新的路由信息下发到数据面使用。若启动防微环路功能，节点在调用 SPF 算法计算得到新的收敛路径后，需要进一步判断收敛路径上是否存在微环路。如果存在，那么节点就会基于 SR 的任意路径编排功能，计算出一条无环 SR 路径段，并将该路径段下发给数据面使用，等到网络重新收敛后，再用收敛路径代替无环路径。这里比较重要的一点就是延长节点从无环路径切换到收敛路径的时间，这个时间可以确保受拓扑变化影响的节点都计算出结果，整个网络重新进入收敛状态。从无环路径切换到收敛路径的时延究竟如何设置，依赖于具体网络场景和经验。设置的主要依据就是确保经过这个时延后，网络能够进入收敛状态。通常，采用 $T = D \cdot \text{Max}_{i=1}^{N}(\text{Delay}_{\text{SPF}(i)})$ 进行时延的计算，D 为网络直径，N 为网络节点数，$\text{Delay}_{\text{SPF}(i)}$ 为节点 i 的路由计算时间。

8.4 计算保护路径的仲裁策略

在进行保护路径计算与选择时涉及两个层面的问题。

一是在计算保护路径前，需要明确被保护的对象是什么。在进行保护路径计算时，首先将被保护对象从当前拓扑上移除，基于剩余链路和节点运行 SPF 算法，将得到的路径作为保护路径。被保护对象可以分为链路、节点、共享风险链路组（Shared Risk Link Group，SRLG）和非共享线卡链路的保护路径。

- 链路的保护路径：针对被保护前缀，从当前拓扑上移除当前节点到最优下一跳节点之间的链路后，计算得到的最优路径为保护路径。
- 节点的保护路径：针对被保护前缀，从当前拓扑上移除当前节点的最优下一跳节点及其所有链路之后，计算得到的最优路径为保护路径。
- SRLG 的保护路径：针对当前节点与邻居节点的特定链路 L，通过预先配置方式，可以将其他链路与链路 L 设置为一个风险共享链路组 G。针对被保护前缀，如果其到最优下一跳的链路属于链路组 G，则从当前拓扑上移除 G 中所有链路，计算得到的最优路径就是保护路径。
- 非共享线卡链路的保护路径：针对被保护前缀，从当前拓扑上移除当前节点到最优下一跳节点之间的链路，进一步移除那些与该链路位于同一个线卡板上的链路，计算得到的最优路径为保护路径。

二是在计算出保护路径后，需要明确指定满足何种条件的路径作为保护路径。默认条件下，优选代价最小的路径作为保护路径，但在实际应用中，可能会因为特殊原因选择满足其他条件的路径。例如，当前 IOS 版本除了最小的路径代价，还提供了优选非 ECMP 路径和 downstream 路径（邻居节点到目的地的链路代价小于本地到目的地的链路代价）的策略。

配置保护路径计算和优选策略的仲裁指令如下：

```
router ospf NAME fast-reroute per-prefix tiebreaker [downstream |
lc-disjoint | lc-disjoint | lowest-backup-metric | node-protecting |
primary-path | secondary-path | srlg-disjoint ] index <1-255>
```

默认计算得到的保护路径为链路保护，并优选最小代价的路径。每种策略

第 8 章 基于 SR 的 FRR 技术原理与实战

之后的 index 值为该策略的优先级，值越大，优先级越高。FRR Tiebreaker 仲裁规则见表 8.1。

表 8.1 FRR Tiebreaker 仲裁规则

仲裁规则	描述
downstream	修复路径入口节点去往目的地的路径代价比 PLR 去往该目的地的路径代价小
lc-disjoint	按照非共享线卡链路方式计算保护路径
lowest-backup-metric	修复路径采用最低度量
node-protecting	按照节点保护方式计算保护路径
primary-path	如果 ECMP 路径存在，则首选该路径作为修复路径
secondary-path	修复路径不能优选 ECMP 路径
srlg-disjoint	按照 SRLG 保护方式计算保护路径

8.4.1 实验环境

仲裁策略实验拓扑如图 8.23 所示，所有节点已按第 2 章的实验约定配置了接口 IP 地址、环回接口地址、MPLS 标签空间等，双向链路代价已标注于图中。所有节点启用 OSPF 协议，都在区域 0 内并且使能 SR 功能。

实验过程中将节点 1 去往节点 5 之间的最优路径作为对象，通过在节点 1 上启动 TI-LFA 路径功能，查看节点 1 上去往目的节点 5 的前缀 5.5.5.5/32 的保护路径。通过改变节点 1 的保护路径仲裁策略，查看在链路保护、节点保护以及 SRLG 保护条件下保护路径的变化情况，从而理解仲裁策略对快速重路由路径的影响。

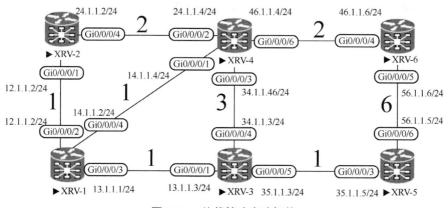

图 8.23 仲裁策略实验拓扑

8.4.2 链路保护

节点启动 TI-LFA 路径后，在没有配置任何仲裁策略的条件下，默认采用链路保护策略来计算保护路径。在节点 1 上启动 TI-LFA 路径功能，查看其去往 5.5.5.5/32 的最优路径、保护路径以及转发路径情况：

```
RP/0/0/CPU0:XRV-1#show ospf routes 5.5.5.5/32 backup-path
Topology Table for ospf 1 with ID 1.1.1.1
O    5.5.5.5/32, metric 3
       13.1.1.3, from 5.5.5.5, via GigabitEthernet0/0/0/3, path-id 1
         Backup path: TI-LFA, Repair-List: P node: 4.4.4.4   Label: 3
                                            Q node: 3.3.3.3  Label: 400000
       14.1.1.4, from 5.5.5.5, via GigabitEthernet0/0/0/4, protected bitmap 0000000000000001
         Attributes: Metric: 6, SRLG Disjoint
RP/0/0/CPU0:XRV-1#show mpls forwarding prefix 5.5.5.5/32  detail
Local  Outgoing   Prefix            Outgoing      Next Hop       Bytes
Label  Label      or ID             Interface                    Switched
------ ---------- ----------------- ------------  ------------   ---
16005  16005      SR Pfx (idx 5)    Gi0/0/0/3     13.1.1.3       0
     Path Flags: 0x400 [ BKUP-IDX:1 (0xa175c9a4) ]
     Version: 125, Priority: 1
     Label Stack (Top -> Bottom): { 16005 }
     NHID: 0x0, Encap-ID: N/A, Path idx: 0, Backup path idx: 1, Weight: 0
     MAC/Encaps: 14/18, MTU: 1500
     Outgoing Interface: GigabitEthernet0/0/0/3 (ifhandle 0x00000080)
     Packets Switched: 0

       Pop        SR Pfx (idx 5)    Gi0/0/0/4     14.1.1.4       0(!)
     Path Flags: 0xb00 [ IDX:1 BKUP, NoFwd ]
     Version: 125, Priority: 1
     Label Stack (Top -> Bottom): { Imp-Null 400000 16005 }
```

```
    NHID: 0x0, Encap-ID: N/A, Path idx: 1, Backup path idx: 0,
Weight: 0
    MAC/Encaps: 14/22, MTU: 1500
    Outgoing Interface: GigabitEthernet0/0/0/4 (ifhandle 0x000000a0)
    Packets Switched: 0
    (!): FRR pure backup
```

可以看到节点 1 去往节点 5 的最优下一跳为节点 3，保护路径为节点 1→4→3→5，即当链路 Link(1,3)失效后，节点 1 对入标签为 16005 的 MPLS 分组采用保护路径转发，具体操作为：弹出栈顶标签，依次压入 16005、400000，然后将 MPLS 分组发送给下一跳(节点 4)。查看节点 4 关于标签 400000 的 MPLS 转发表项：

```
RP/0/0/CPU0:XRV-4#show mpls forwarding labels 400000 detail
Local    Outgoing   Prefix            Outgoing        Next Hop       Bytes
Label    Label      or ID             Interface                      Switched
------   --------   ---------------   -------------   -----------    --------
400000   Pop        SR Adj (idx 0)    Gi0/0/0/3       34.1.1.3       0
Version: 5, Priority: 1
Label Stack (Top -> Bottom): { Imp-Null }
NHID: 0x0, Encap-ID: N/A, Path idx: 0, Backup path idx: 0, Weight: 0
MAC/Encaps: 14/14, MTU: 1500
Outgoing Interface: GigabitEthernet0/0/0/3 (ifhandle 0x00000080)
Packets Switched: 0
```

可见，节点 4 收到标签为 400000 的 MPLS 分组后，将该标签弹出，通过接口 Gi0/0/0/3 转发给下一跳节点 34.1.1.3，即节点 3。到达节点 3 的 MPLS 分组栈顶标签为 16005，节点 3 弹出该标签后转发给节点 5，从而实现了对链路故障时数据传送的保护。

8.4.3 节点保护

对于节点 1 而言，除了 Link(1,3)故障，还有可能出现最优下一跳（节点 3）故障的现象。如果采用链路保护策略，计算得到的保护路径为节点

1→4→3→5，可见保护路径经过了故障节点，因此，该保护路径无法实现对节点失效的保护。

为了解决这个问题，可以在节点 1 上使能节点保护功能，让该节点按照节点保护策略进行保护路径的计算。在节点 1 上配置节点保护指令后，查看保护路径信息：

```
RP/0/0/CPU0:XRV-1(config)#router ospf 1 fast-reroute per-prefix tie breaker node-protecting index 100
RP/0/0/CPU0:XRV-1#show ospf routes 5.5.5.5/32 backup-path
Topology Table for ospf 1 with ID 1.1.1.1
O    5.5.5.5/32, metric 3
       13.1.1.3, from 5.5.5.5, via GigabitEthernet0/0/0/3, path-id 1
         Backup path: TI-LFA, Repair-List: P node: 6.6.6.6    Label: 16006
                                            Q node: 5.5.5.5   Label: 600000
       14.1.1.4, from 5.5.5.5, via GigabitEthernet0/0/0/4, protected bitmap 0000000000000001
         Attributes: Metric: 10, Node Protect, SRLG Disjoint
RP/0/0/CPU0:XRV-1#show mpls forwarding prefix 5.5.5.5/32  detail
Local  Outgoing    Prefix           Outgoing      Next Hop     Bytes
Label  Label       or ID            Interface                  Switched
------ ----------- ---------------- ------------ ------------ ------
16005  16005       SR Pfx (idx 5)   Gi0/0/0/3    13.1.1.3     0
    Path Flags: 0x400 [ BKUP-IDX:1 (0xa175c65c) ]
    Version: 136, Priority: 1
    Label Stack (Top -> Bottom): { 16005 }
    NHID: 0x0, Encap-ID: N/A, Path idx: 0, Backup path idx: 1, Weight: 0
    MAC/Encaps: 14/18, MTU: 1500
    Outgoing Interface: GigabitEthernet0/0/0/3 (ifhandle 0x00000080)
    Packets Switched: 0

       16006       SR Pfx (idx 5)   Gi0/0/0/4    14.1.1.4     0(!)
```

```
    Updated: Apr  3 05:23:38.044
    Path Flags: 0xb00 [ IDX:1 BKUP, NoFwd ]
    Version: 136, Priority: 1
    Label Stack (Top -> Bottom): { 16006 600000 }
    NHID: 0x0, Encap-ID: N/A, Path idx: 1, Backup path idx: 0,
Weight: 0
    MAC/Encaps: 14/22, MTU: 1500
    Outgoing Interface: GigabitEthernet0/0/0/4 (ifhandle 0x000000a0)
    Packets Switched: 0
    (!): FRR pure backup
```

此时节点1计算得出的保护路径为节点1→4→6→5，该路径避开了节点3，实现了对节点3失效的故障保护。从数据面来说，携带16005标签的分组到达节点1时，如果遭遇链路故障，则弹出栈顶标签16005，然后依次压入标签600000、16006，形成保护数据分组，然后通过接口Gi0/0/0/4转发给节点4。

查看节点4关于标签16006的MPLS转发表项：

```
RP/0/0/CPU0:XRV-4#show mpls forwarding labels 16006 detail
Local    Outgoing    Prefix            Outgoing       Next Hop     Bytes
Label    Label       or ID             Interface                   Switched
------   ---------   ---------------   ------------   -----------  --------
16006    Pop         SR Pfx (idx 6)    Gi0/0/0/6      46.1.1.6     0
    Updated: Apr  3 03:18:33.898
    Version: 51, Priority: 1
    Label Stack (Top -> Bottom): { Imp-Null }
    NHID: 0x0, Encap-ID: N/A, Path idx: 0, Backup path idx: 0,
Weight: 0
    MAC/Encaps: 14/14, MTU: 1500
    Outgoing Interface: GigabitEthernet0/0/0/6 (ifhandle 0x000000e0)
    Packets Switched: 0
```

可见，节点4将保护数据分组栈顶标签16006弹出后通过接口Gi0/0/0/6转发给节点6。此时保护数据分组栈顶标签变为600000，该分组到达节点6后，节点6弹出该标签，还原为原始分组，通过Gi0/0/0/5接口转发给节点5，从而实现了对节点3失效的故障保护。

8.4.4 SRLG 保护

在 MPLS 流量工程中，共享风险链路组（SRLG）是一组共享公共资源的链路，如果公共资源发生故障，会影响其中的所有链路。这些链路具有相同的故障风险，因此被视为属于相同的 SRLG。

SRLG 由 IGP（OSPFv2 或 IS-IS）域内唯一的 32 位数字表示。一个链路可能属于多个 SRLG。在计算备用路径时，优先选择和主用路径不在同一 SRLG 的备用路径。这样可以保证特定链路上的单点故障不会同时影响主用路径和备用路径。

将链路设置为同一个 SRLG 的指令为：

```
srlg group <GroupName> value <0-4294967295>//创建组
srlg interface <IFNAME> group 1 <GroupName>//将接口绑定到组
```

在节点 1 上将 Gi0/0/0/3 和 Gi0/0/0/4 设置为一个 SRLG 组，节点 1 的 SRLG 配置信息为：

```
srlg
 interface GigabitEthernet0/0/0/3
  group
   1 G2
  !
 !
 interface GigabitEthernet0/0/0/4
  group
   1 G2
  !
 !
 group G2
  8 value 2000
 !
!
```

在设置了 SRLG 后，配置节点 1 优先使用 SRLG 策略计算保护路径，为了避免混淆，去掉之前配置的节点保护策略，并查看节点 1 计算得到的保护路径与转发路径信息：

```
RP/0/0/CPU0:XRV-1(config)#no router ospf 1 fast-reroute per-prefix
tiebreaker node-protecting index 100
RP/0/0/CPU0:XRV-1(config)#router ospf 1 fast-reroute per-prefix tie
breaker srlg-disjoint index 150
RP/0/0/CPU0:XRV-1#show ospf routes 5.5.5.5/32 backup-path
Topology Table for ospf 1 with ID 1.1.1.1
O    5.5.5.5/32, metric 3
        13.1.1.3, from 5.5.5.5, via GigabitEthernet0/0/0/3, path-id 1
          Backup path: TI-LFA, Repair-List:  P node: 2.2.2.2  Label: 3
                                             Q node: 4.4.4.4  Label:
200001
                                             Q node: 3.3.3.3  Label:
400000
           12.1.1.2, from 5.5.5.5, via GigabitEthernet0/0/0/2, pro
tected bitmap 0000000000000001
           Attributes: Metric: 8, SRLG Disjoint
RP/0/0/CPU0:XRV-1#show mpls forwarding prefix 5.5.5.5/32 detail
Local   Outgoing  Prefix          Outgoing     Next Hop      Bytes
Label   Label     or ID           Interface                  Switched
------  --------  --------------  -----------  ------------  --------
16005   1600      SR Pfx (idx 5)  Gi0/0/0/3    13.1.1.3      0
    Path Flags: 0x400 [ BKUP-IDX:0 (0xa175c918) ]
    Version: 179, Priority: 1
    Label Stack (Top -> Bottom): { 16005 }
    NHID: 0x0, Encap-ID: N/A, Path idx: 1, Backup path idx: 0,
Weight: 0
    MAC/Encaps: 14/18, MTU: 1500
    Outgoing Interface: GigabitEthernet0/0/0/3 (ifhandle 0x00000080)
    Packets Switched: 0
```

```
  200001       SR Pfx (idx 5)    Gi0/0/0/2   12.1.1.2        0(!)
     Path Flags: 0xb00 [ IDX:0 BKUP, NoFwd ]
     Version: 179, Priority: 1
     Label Stack (Top -> Bottom): { 200001 400000 16005 }
     NHID: 0x0, Encap-ID: N/A, Path idx: 0, Backup path idx: 0,
Weight: 0
     MAC/Encaps: 14/26, MTU: 1500
     Outgoing Interface: GigabitEthernet0/0/0/2 (ifhandle 0x00000060)
     Packets Switched: 0
     (!): FRR pure backup
```

可以看到，由于节点1将通向节点3和节点4的链路设置为一个链路组，因此，节点 1 将这两条链路从拓扑上删除后计算得到的保护路径为节点 1→2→4→3→5。

从数据面来说，携带16005标签的数据分组到达节点1时，如果遭遇链路故障，则弹出栈顶标签后，依次压入标签 16005、400000、200001，形成保护数据分组，然后通过接口 Gi0/0/0/2 转发给节点 2。节点 2 将保护数据分组栈顶标签 200001 弹出后转发给节点 4，节点 4 弹出 400000 标签后转发给节点 3，节点 3 弹出 16005 标签后将分组转发给节点 5，从而实现了数据转发保护，保护路径成功绕开了 Link(1,3)和 Link(1,4)。

8.4.5　Node+SRLG 保护

上文仅阐述了链路保护、节点保护以及 SRLG 保护等单一策略生效情况下的保护路径计算。下面在 SRLG 保护场景的基础上，通过配置节点保护策略，探讨多种策略并存条件下保护路径的计算。

在节点 1 上同时配置节点保护和 SRLG 保护，Gi0/0/0/3 和 Gi0/0/0/4 属于同一个 SRLG 组，此时节点 1 的配置信息如下：

```
router ospf 1
 segment-routing mpls
 segment-routing forwarding mpls
```

```
fast-reroute per-prefix
fast-reroute per-prefix ti-lfa enable
segment-routing sr-prefer
fast-reroute per-prefix tiebreaker node-protecting index 100
fast-reroute per-prefix tiebreaker srlg-disjoint index 150
```

查看节点1计算得到的保护路径与转发路径信息：

```
RP/0/0/CPU0:XRV-1#show ospf routes 5.5.5.5/32 backup-path
Topology Table for ospf 1 with ID 1.1.1.1
O    5.5.5.5/32, metric 3
       13.1.1.3, from 5.5.5.5, via GigabitEthernet0/0/0/3, path-id 1
         Backup path: TI-LFA, Repair-List:  P node: 2.2.2.2  Label: 3
                                            Q node: 4.4.4.4  Label: 200001
                                            P node: 6.6.6.6  Label: 16006
                                            Q node: 5.5.5.5  Label: 600000
       12.1.1.2, from 5.5.5.5, via GigabitEthernet0/0/0/2,
protected bitmap 0000000000000001
           Attributes: Metric: 12, Node Protect, SRLG Disjoint
RP/0/0/CPU0:XRV-1#show mpls forwarding prefix 5.5.5.5/32  detail
Local  Outgoing    Prefix           Outgoing      Next Hop     Byte
Label  Label       or ID            Interface                  Switched
------ ----------- ---------------- ------------- ------------ ---------
16005  16005       SR Pfx (idx 5)   Gi0/0/0/3     13.1.1.3     0
    Path Flags: 0x400 [ BKUP-IDX:0 (0xa175c918) ]
    Version: 169, Priority: 1
    Label Stack (Top -> Bottom): { 16005 }
    NHID: 0x0, Encap-ID: N/A, Path idx: 1, Backup path idx: 0, Wei
ght: 0
    MAC/Encaps: 14/18, MTU: 1500
    Outgoing Interface: GigabitEthernet0/0/0/3 (ifhandle 0x00000080)
    Packets Switched: 0

200001        SR Pfx (idx 5)   Gi0/0/0/2     12.1.1.2     0(!)
    Path Flags: 0xb00 [ IDX:0 BKUP, NoFwd ]
```

```
    Version: 169, Priority: 1
    Label Stack (Top -> Bottom): { 200001 16006 600000 }
    NHID: 0x0, Encap-ID: N/A, Path idx: 0, Backup path idx: 0, Weight: 0
    MAC/Encaps: 14/26, MTU: 1500
    Outgoing Interface: GigabitEthernet0/0/0/2 (ifhandle 0x00000060)
    Packets Switched: 0
    (!): FRR pure backup
```

从 OSPF 进程下显示的备用路径来看，当前保护路径能够同时满足"Node Protect"与"SRLG Disjoint"两个条件。计算得到的保护路径为节点 1→2→4→6→5，保护路径既避开了 SRLG 内的链路，也避开了节点 3，从而同时满足了节点保护和 SRLG 保护条件。

从数据面来说，携带16005标签的数据分组到达节点1时，如果遭遇故障，则弹出栈顶标签后，依次压入标签 600000、16006、200001，形成保护数据分组，然后通过接口 Gi0/0/0/2 转发给节点 2。标签 200001 指导节点 2 将保护流量转发给节点 4，16006 指导节点 4 将流量转发给节点 6，600000 指导节点 6 将流量转发给节点 5，实现故障时流量保护。

在保持上述配置不变的条件下，将节点 1 的 Gi0/0/0/2 接口也加入 SRLG 中，查看节点的保护路径计算情况：

```
RP/0/0/CPU0:XRV-1#show ospf routes 5.5.5.5/32 backup-path
Topology Table for ospf 1 with ID 1.1.1.1
O    5.5.5.5/32, metric 3
        13.1.1.3, from 5.5.5.5, via GigabitEthernet0/0/0/3, path-id 1
        Backup path: TI-LFA, Repair-List: P node: 6.6.6.6  Label: 16006
                                          Q node: 5.5.5.5  Label: 600000
            14.1.1.4, from 5.5.5.5, via GigabitEthernet0/0/0/4,
protected bitmap 0000000000000001
                Attributes: Metric: 10, Node Protect,
```

可以看到，此时基于 SRLG 的保护策略无法计算得到保护路径，但是保护路径还是存在的，只不过此时的保护路径仅仅满足节点保护策略，而无法满足 SRLG 保护策略。